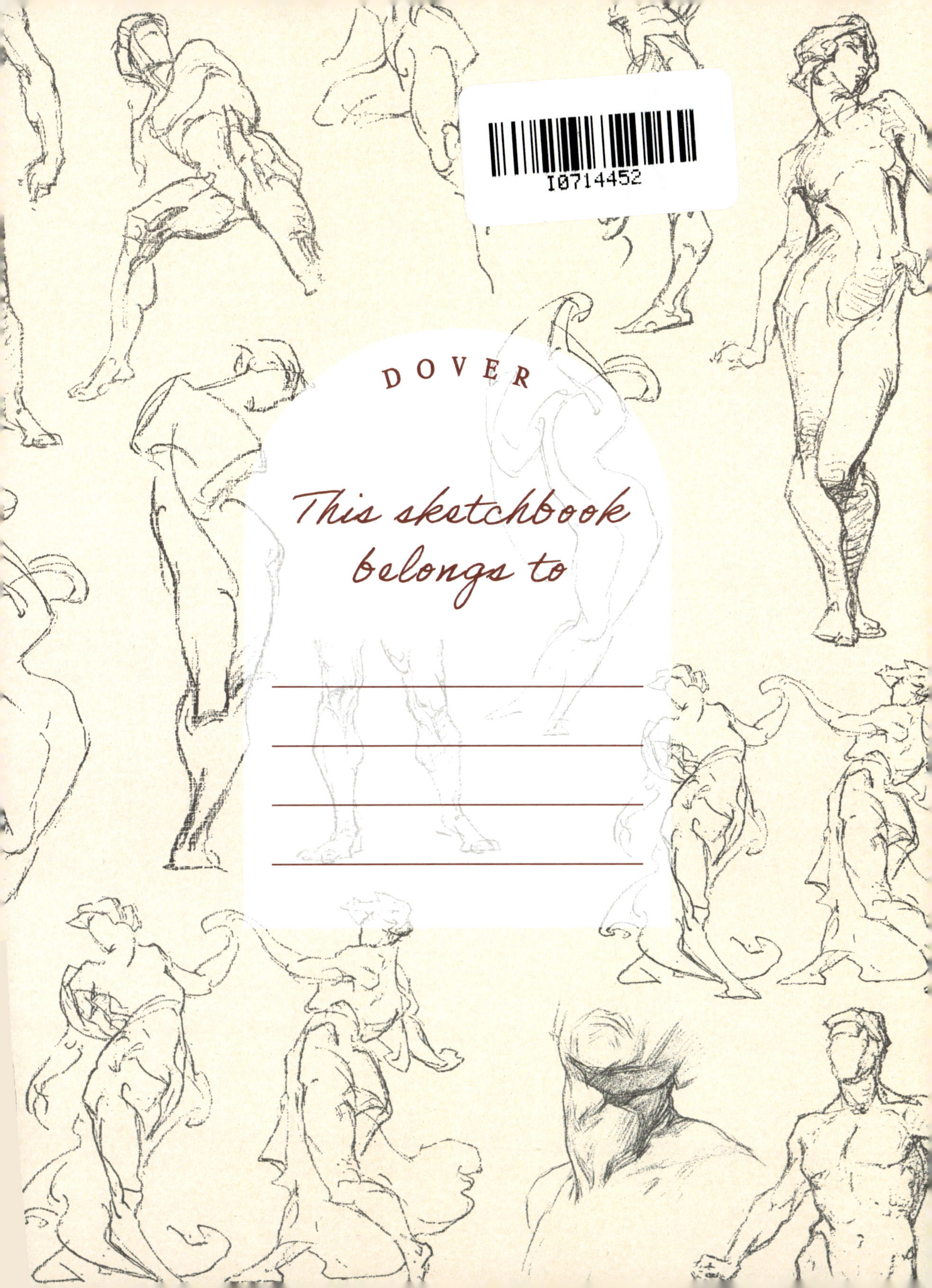

DOVER
This sketchbook
belongs to

Classical Drawing: The Figure: A Guided Sketchbook with Prompts & Tips from the Pros
is a new work, first published by Dover Publications in 2025.

ISBN-13: 978-0-486-85407-6
ISBN-10: 0-486-85407-8

Publisher: Betina Cochran
Acquisitions Editor: Allyson D'Antonio
Managing Editorial Supervisor: Susan Rattiner
Production Editor: Gregory Koutrouby
Editorial, Design, and Layout: Elizabeth T. Gilbert and Coffee Cup Creative LLC
Creative Manager: Marie Zaczkiewicz
Production: Pam Weston, Tammi McKenna, Ayse Yilmaz

Printed in China
85407801   2025
www.doverpublications.com

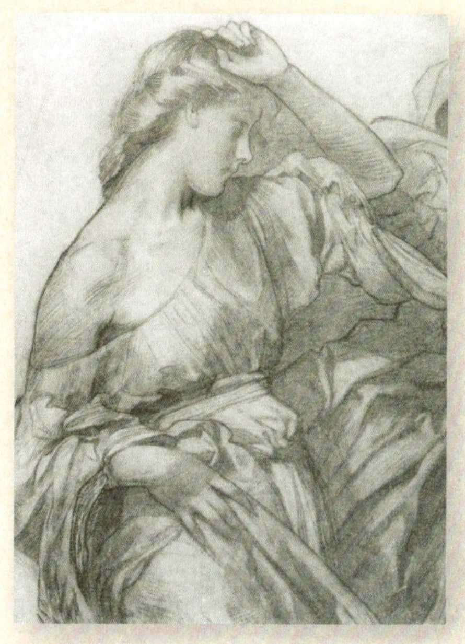

CLASSICAL DRAWING

The Figure

A GUIDED SKETCHBOOK
WITH PROMPTS & TIPS
FROM THE PROS

George Bridgman & Harold Speed

Dover Publications
Garden City, New York

TABLE OF CONTENTS

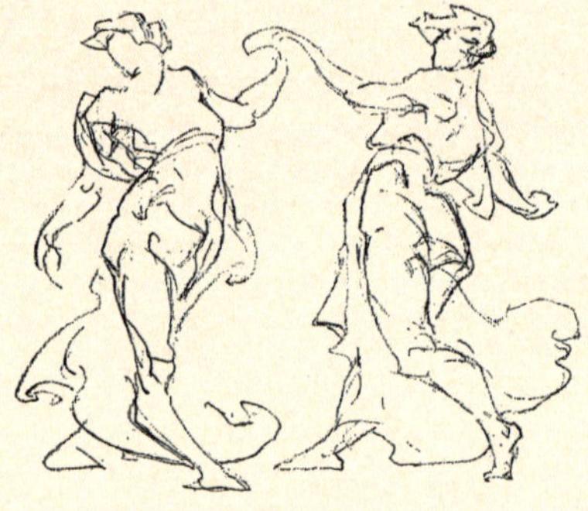

PAGE

Introduction 5

Getting Started 7

Tools & Materials 8

Warming Up 12

Anatomy 16

Proportion 20

Form 24

Balance 32

Movement 42

The Torso 54

Arms & Hands 64

Legs & Feet 70

The Head 78

Sketches & Notes 95

INTRODUCTION

Figures are one of the most complex subjects in art, calling for careful attention to proportion, anatomy, and form. And unlike other subjects, artists must capture movement, a sense of balance, and fleeting human expression. Cultivating this ability takes plenty of time and practice. A sketchbook is an ideal place to begin studying this art form. You'll have the freedom to experiment without consequence, and you can record as many notes and studies as you wish. Frequent sketching develops your observational skills and helps you build on your knowledge so you can increase your speed and accuracy while honing your personal style.

This guided sketchbook features classic sketches from renowned artists George Bridgman and Harold Speed to inspire and instruct you as you dive into the art of sketching figures. Simply sketch alongside the studies in this book, and use the pages at the end for recording your own original sketches and notes.

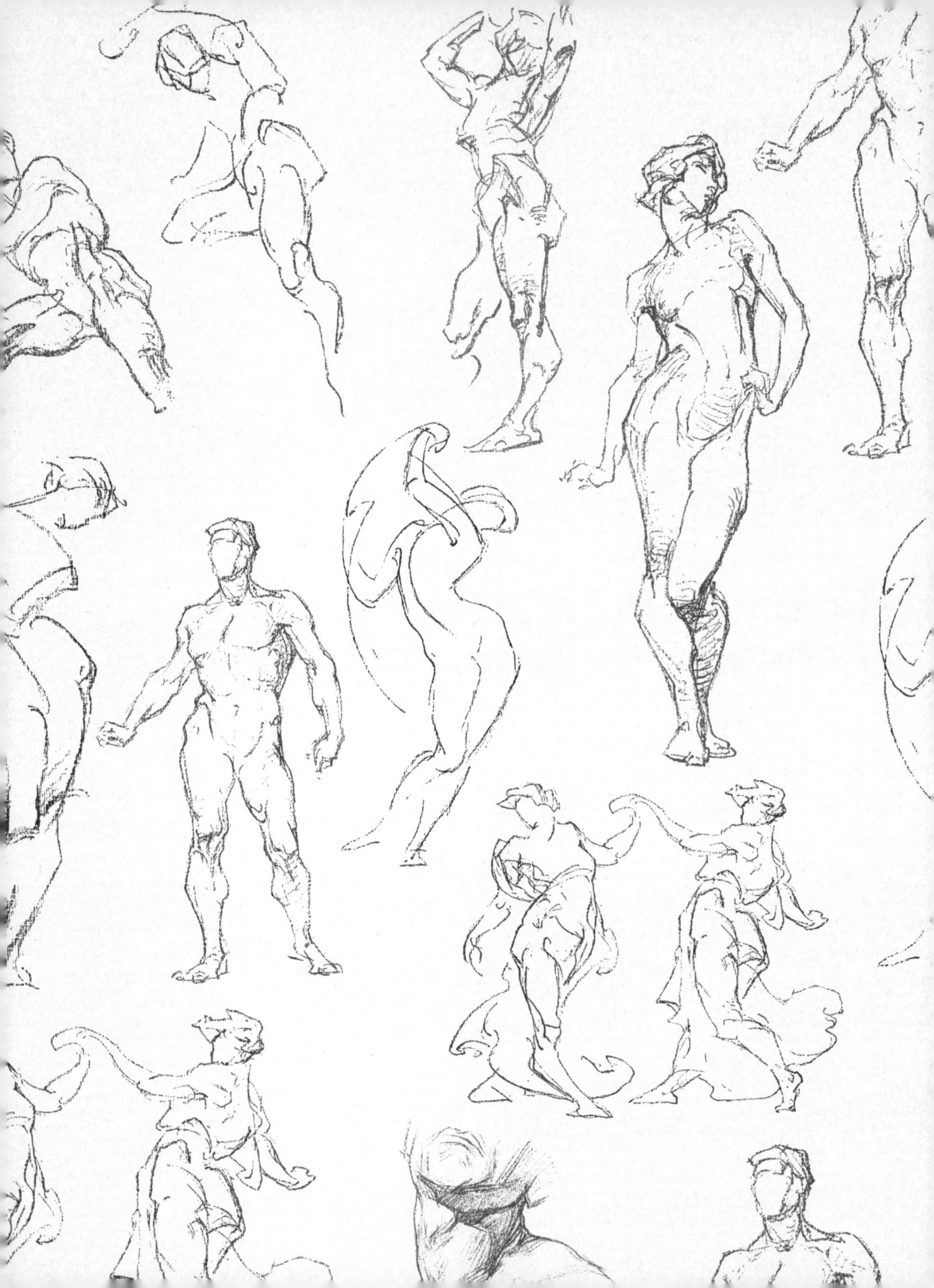

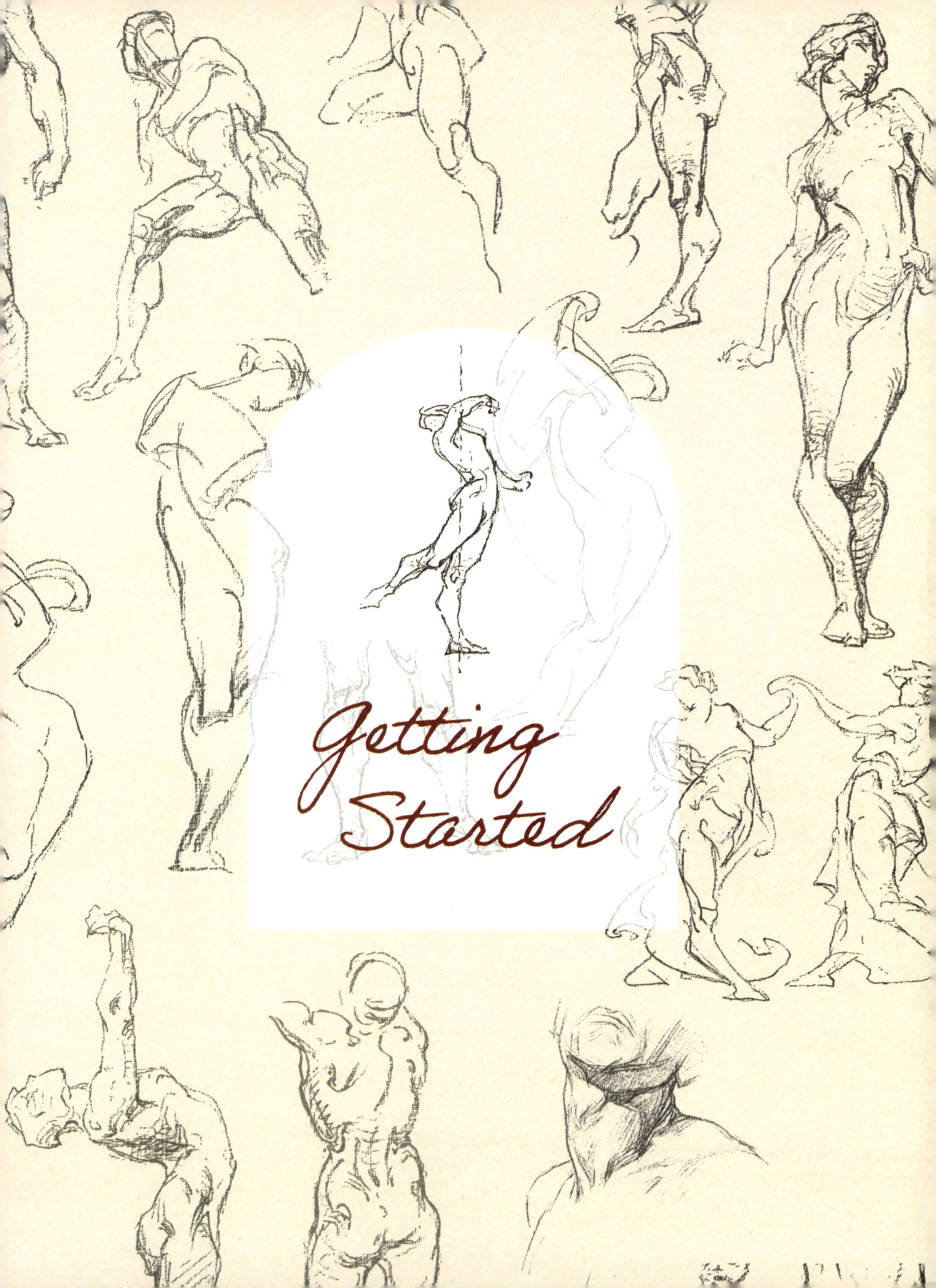

Getting Started

TOOLS & MATERIALS

Pencil drawing is one of the easiest fine arts to master. It requires only a few tools, and cleanup is simple. Moreover, with the help of a small bag or case, pencil drawing is easy to perform on the go and in the presence of your subject. The following pages will help you gather the tools you'll need to get the most out of this sketchbook.

Graphite Pencils

Drawing pencils are generally wood-encased pencils with a graphite lead. Leads are labeled by hardness, ranging from very soft (6B) to medium (HB) to very hard (6H). For sketching figures, it's a good idea to begin simply with a few soft pencils, such as a 2B and a 4B. These will help you create darker tones while moving quickly and loosely.

Sharpeners

Keep a pencil sharpener on hand for maintaining sharp tips. Sharpeners with a small compartment for collecting the shavings are convenient for artists on the go. If you prefer sketching with flat or blunt tips, you can further hone the lead tip using fine sandpaper or a crafting knife.

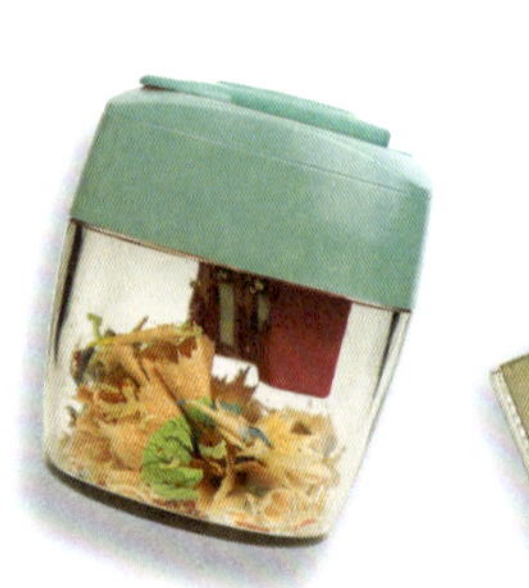

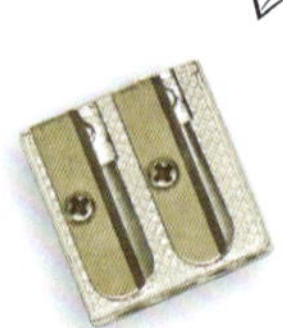

Erasers

Rubber (A), plastic, gum, and kneaded erasers are all effective in removing graphite from paper. Kneaded erasers (B) are great art tools in themselves. You can use them to gently dab away graphite and subtly lighten tones, or you can shape them into fine points to create detailed highlights.

A

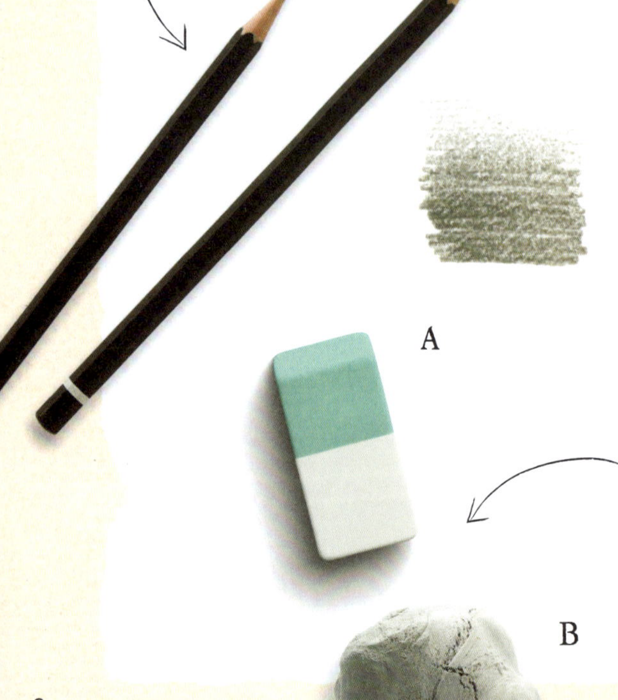

B

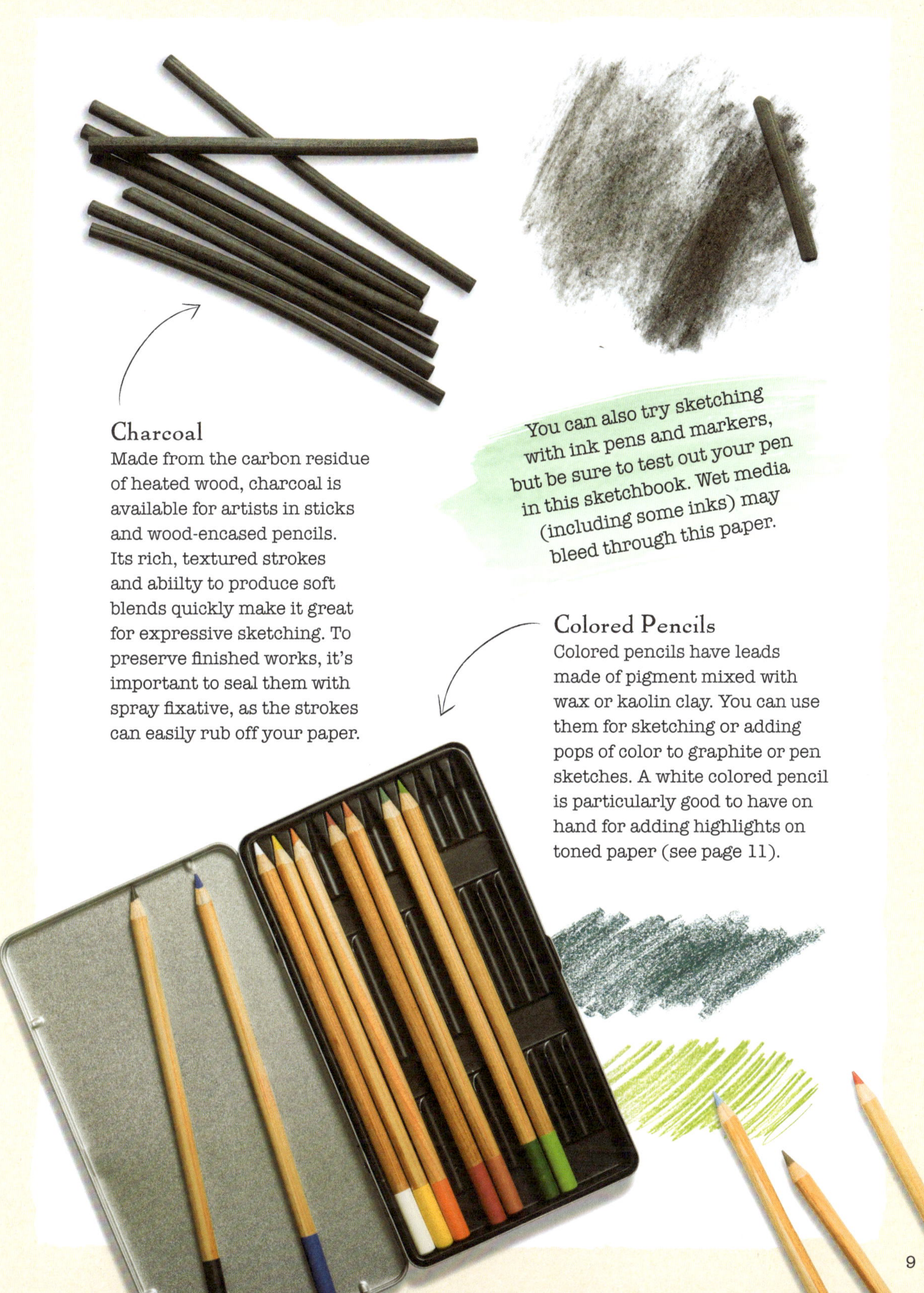

Charcoal

Made from the carbon residue of heated wood, charcoal is available for artists in sticks and wood-encased pencils. Its rich, textured strokes and abiilty to produce soft blends quickly make it great for expressive sketching. To preserve finished works, it's important to seal them with spray fixative, as the strokes can easily rub off your paper.

You can also try sketching with ink pens and markers, but be sure to test out your pen in this sketchbook. Wet media (including some inks) may bleed through this paper.

Colored Pencils

Colored pencils have leads made of pigment mixed with wax or kaolin clay. You can use them for sketching or adding pops of color to graphite or pen sketches. A white colored pencil is particularly good to have on hand for adding highlights on toned paper (see page 11).

Sketchbooks

Once you've filled up this book with drawings, you'll want to purchase a sketchbook. These portable drawing pads come in a range of sizes, bindings, and paper types. Choose a format that feels comfortable to you and a paper type that is suitable for your drawing media of choice. Smooth or medium-textured paper is best for graphite, whereas rough or grain-textured paper is ideal for charcoal. A mixed-media sketchbook is a great choice for beginners and experimenters, as this versatile paper will be heavy enough to accept ink.

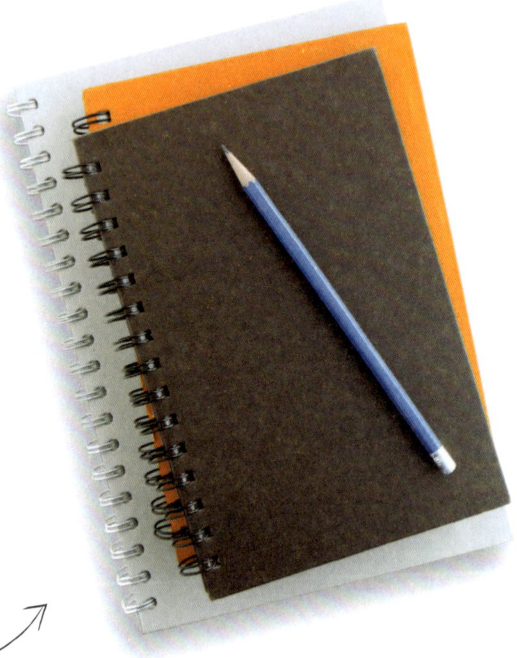

Paper

Many artists use their sketchbooks to create preliminary drawings for more polished works of art. For finished art, choose a high-quality, heavy-weight paper with a texture that suits your drawing style. Consider using spray fixative to set the artwork and prevent your strokes from smudging.

Working on Toned Paper

Sketching on toned paper is a common approach in classical drawing.
In this method, the tone of the paper (generally gray or a neutral color)
serves as a midtone. The artist builds form by adding the shadows and
light. If working on gray paper, the artist could work with soft graphite,
black colored pencil, or charcoal, along with a white pencil.

Some artists choose to tone their own white drawing paper by blending
graphite or charcoal over the surface with a tissue or chamois cloth. Then
they use erasers—such as a stick or kneaded eraser—to pull out tone
for light areas and highlights. White pencil can also be used to create
more dramatic highlights.

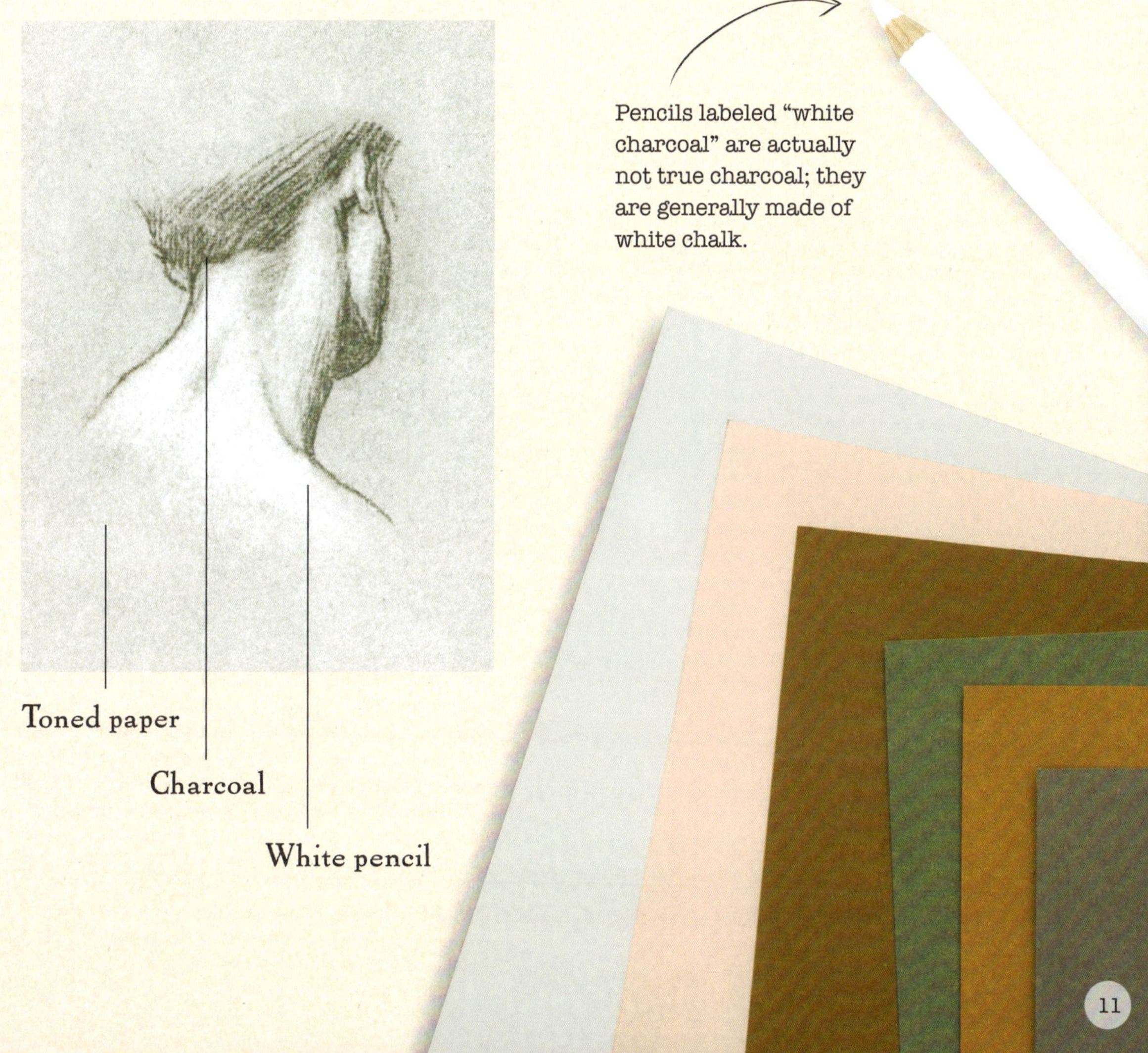

Pencils labeled "white charcoal" are actually not true charcoal; they are generally made of white chalk.

WARMING UP

Before you begin sketching, take some time to warm up and get
comfortable with your drawing tools. Perform some simple strokes and
exercises to loosen up your wrist, develop control over the pencil, and
practice long, broad strokes for shading. The following pages provide
samples to guide you through the warm-up process.

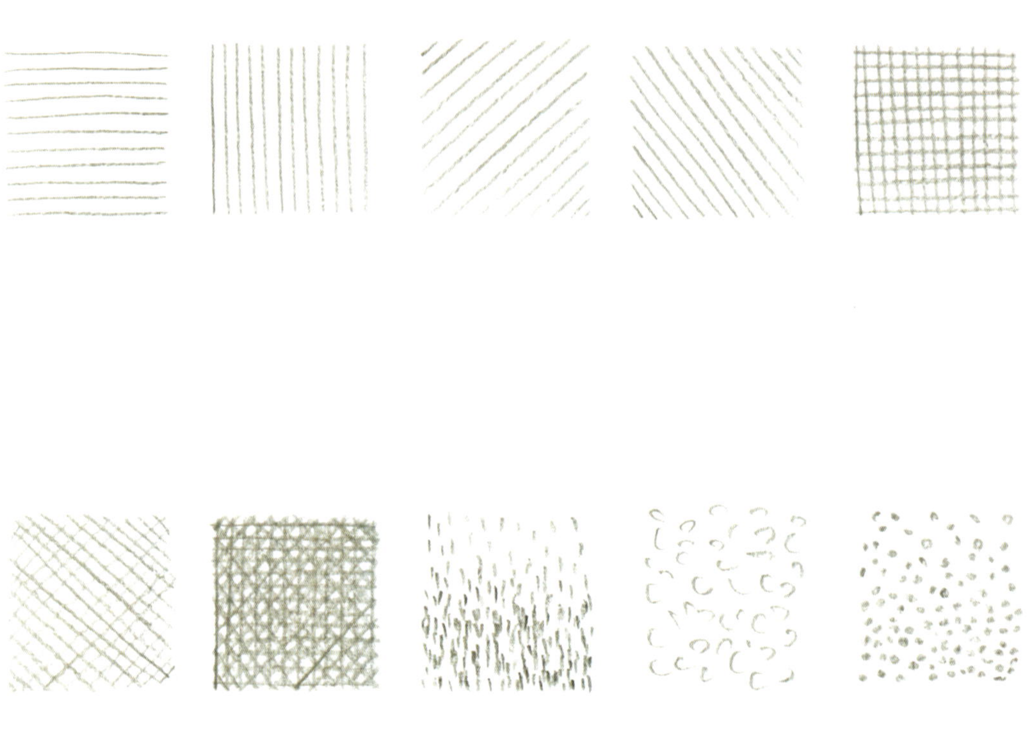

Draw along in this book under each example, or use a separate sheet of scrap paper.
You'll be using a variety of marks to render figures, so think of this exercise as expanding
your "vocabulary" of strokes.

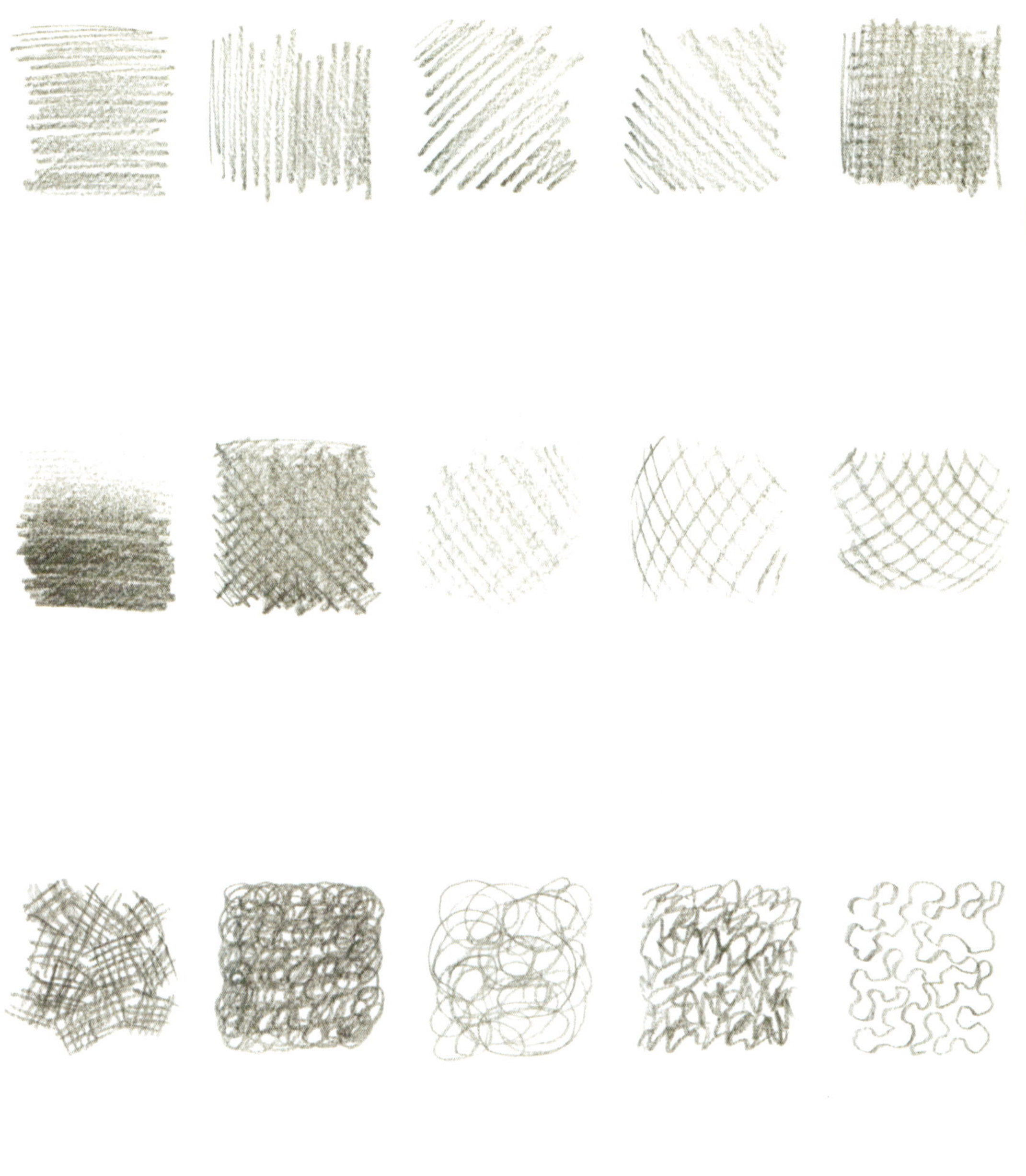

While warming up, experiment with different pencil grips. An overhand grip—lightly pinching the pencil between the thumb and forefinger—will help you create long, sweeping, expressive strokes. A writing grip is best for finer details.

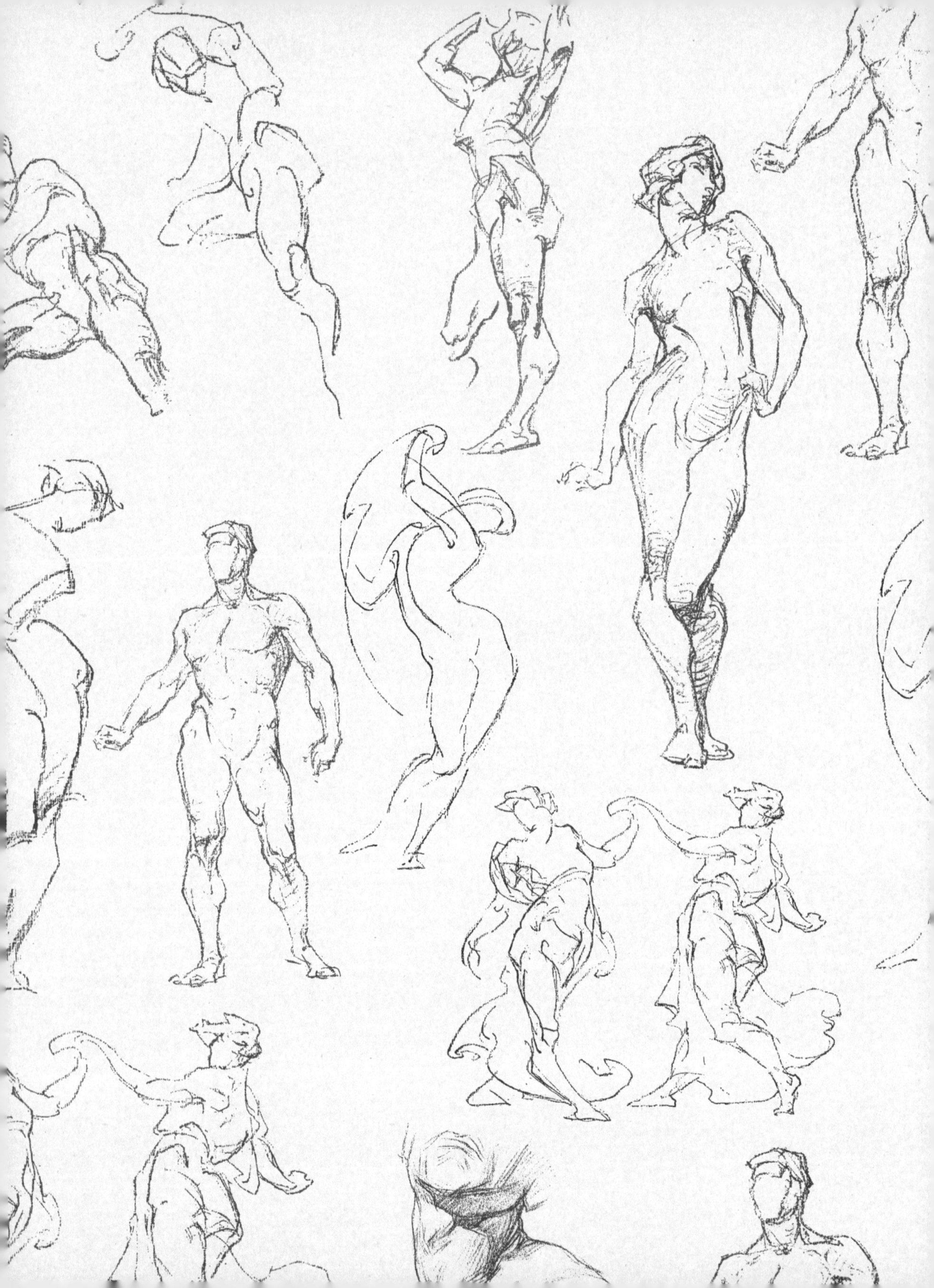

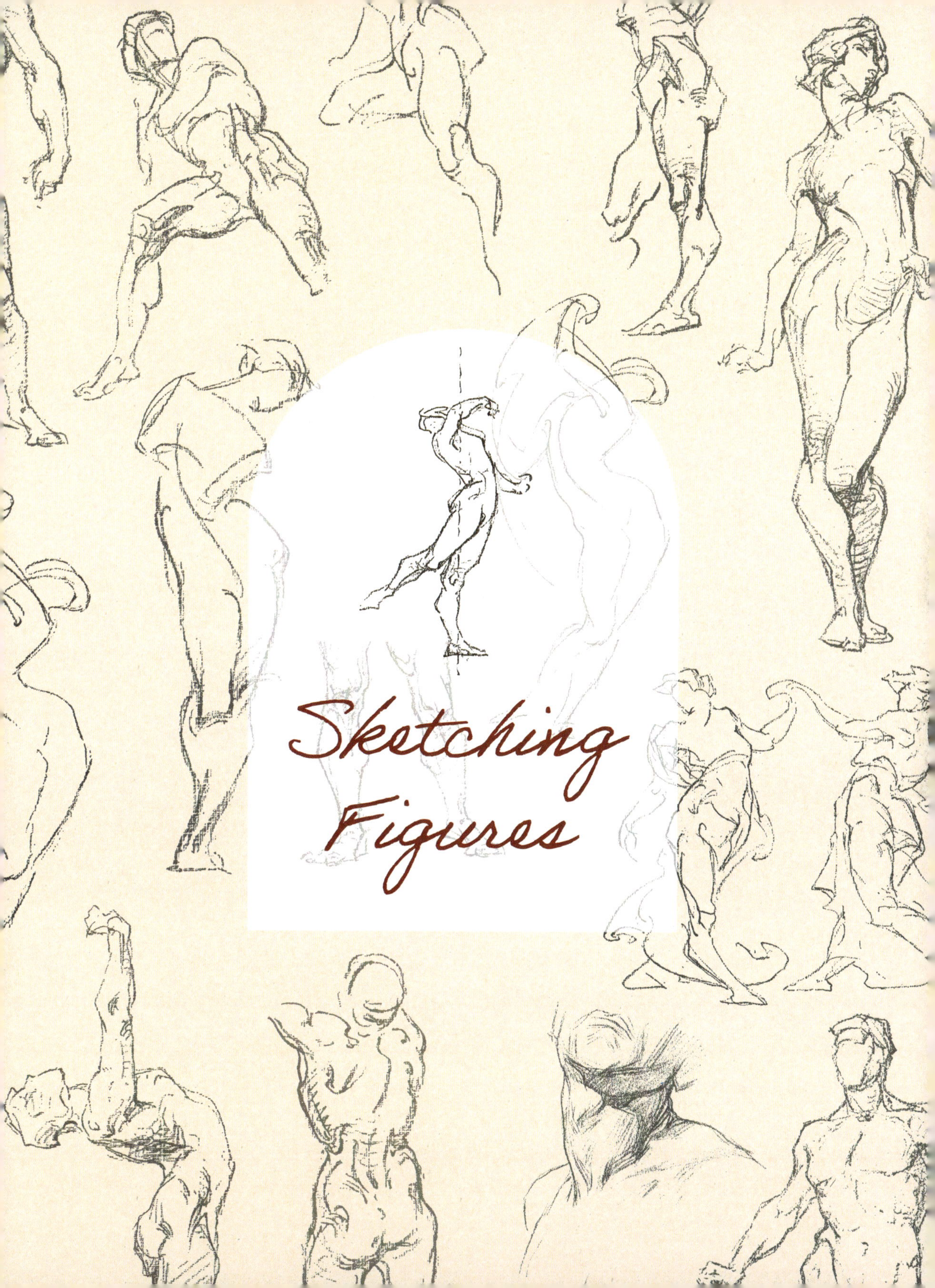
Sketching
Figures

ANATOMY

Human anatomy is the study of the body's structures. Learning the basics of the human skeleton is an excellent way to understand how the underlying structures influence the forms of the body and the way we move. Study the anatomy sketches on this page and on page 18. Then use the blank pages to practice.

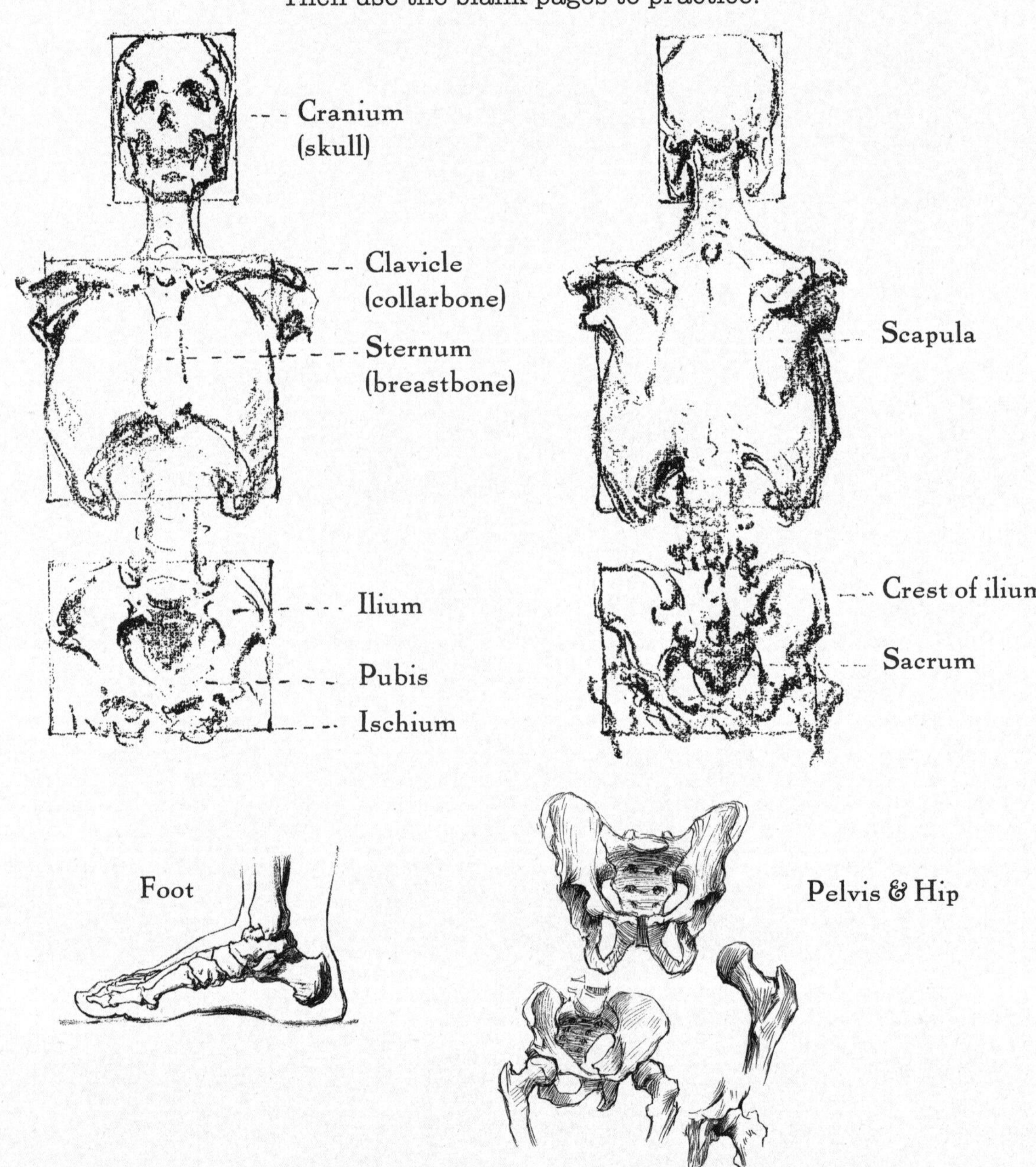

1. Frontal

2. Temporal

3. Zygomatic arch

4. Malar

5. Superior maxillary

6. Inferior maxillary

7. Nasal

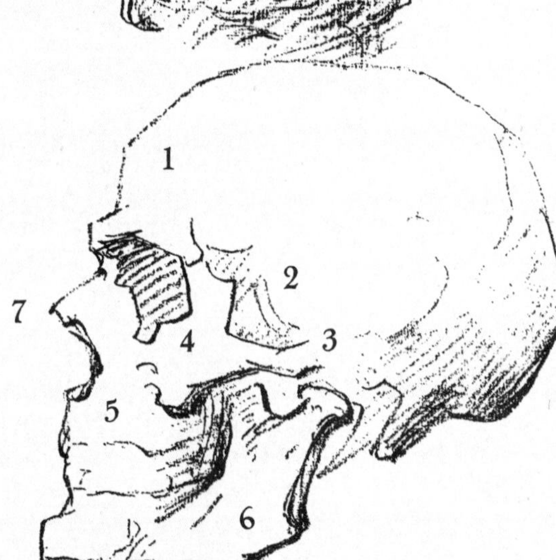

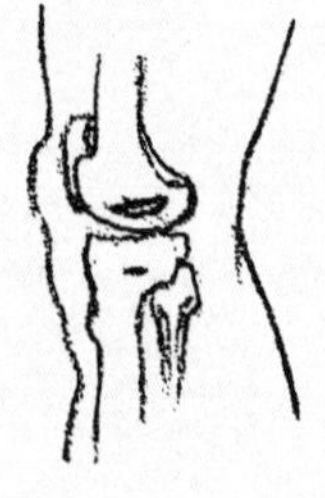

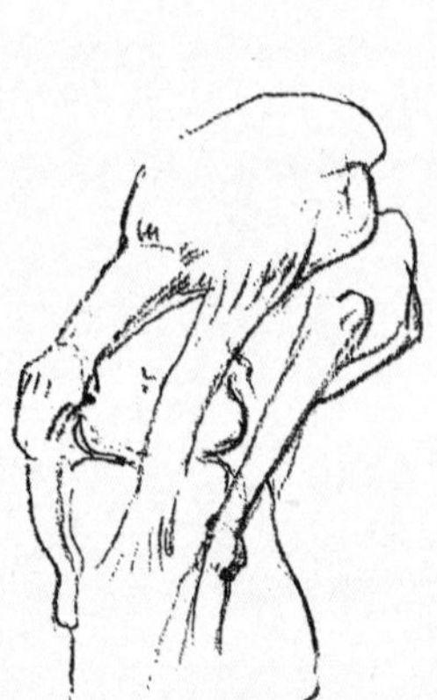

PROPORTION

The human figure can be measured using divisions of the body into parts. How these parts relate to each other (and to the whole body) will provide you with general human proportions. The visuals below show proportion studies applied to an upright figure standing at eye level. As every body is unique and varies somewhat, think of these methods of measurement as a loose guide.

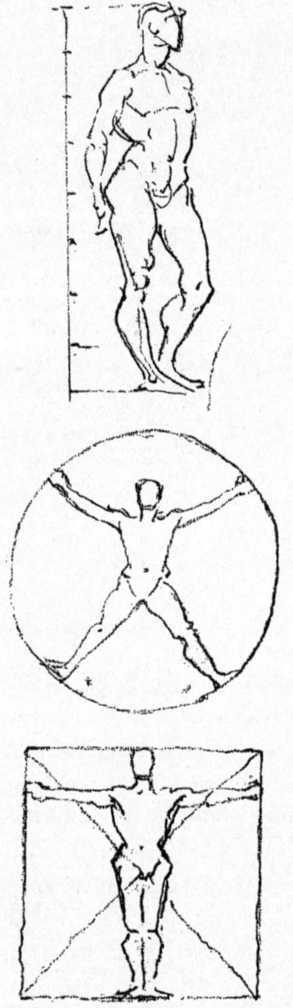

One of the most revered artists of the Renaissance, Michelangelo based his works on the ideal human body as about 8 heads tall.

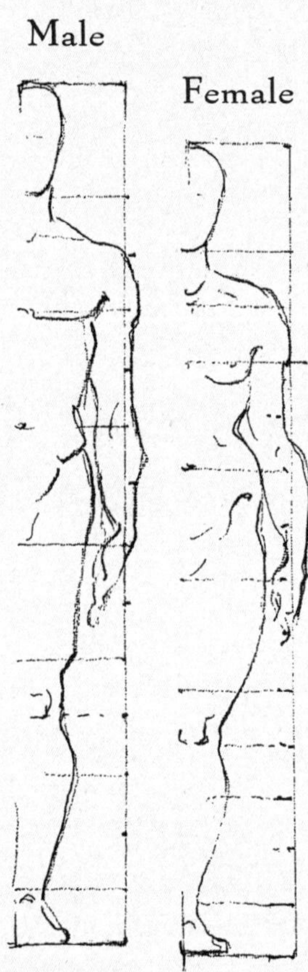

French scientist, doctor, and artist Dr. Paul Richer (1849–1933) described the human body as about 7 1/2 heads tall.

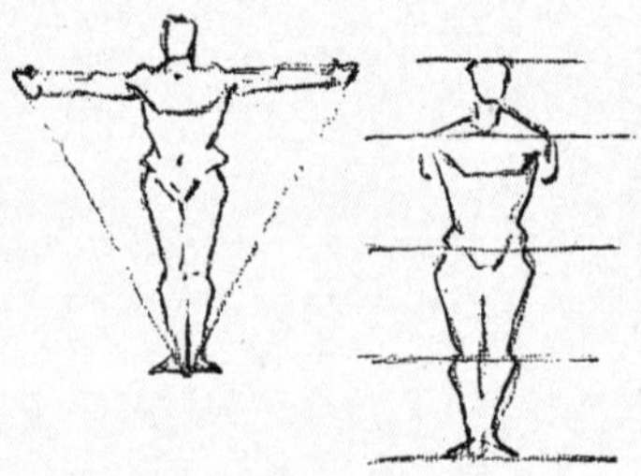

The American anatomist and artist Dr. William Rimmer (1816–1879) is known for his extensive studies on human proportions, as well as the skeletal and muscular structures that define the body's form.

Building the Body
The skull is often used as a unit of measurement for the body. Taking the skull as a horizontal unit, the bone of the upper arm (humerus) is about 1 1/2 heads in length. The bone on the thumb side of the forearm (radius) is about 1 head in length. The forearm bone (ulna) on the little finger side measures about 1 head from elbow to wrist. The thigh bone (femur) measures about 2 heads, and the leg bone (tibia) measures nearly 1 1/2 heads.

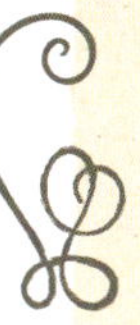

Use the practice pages in this section to create human proportion studies for future reference.

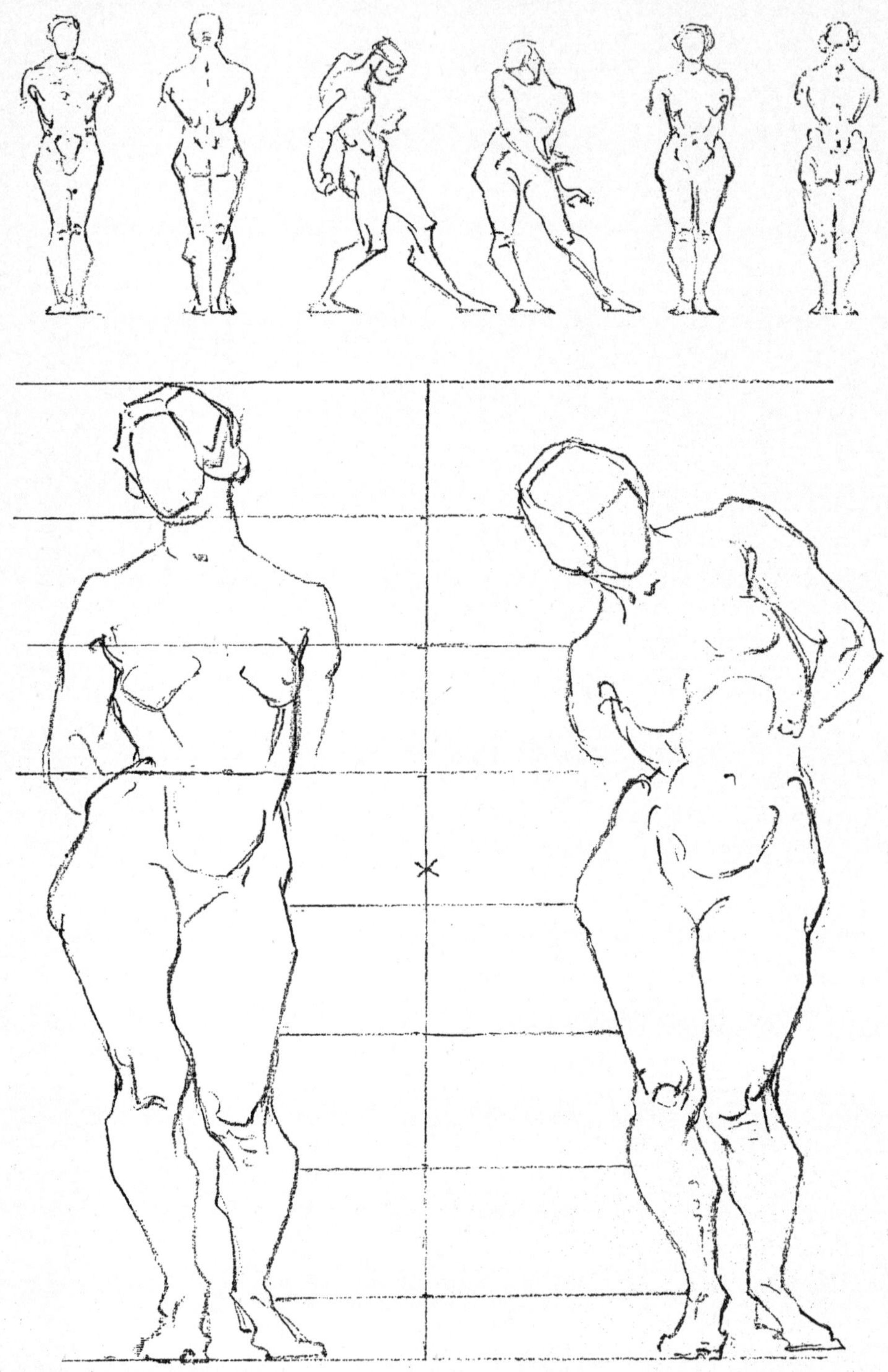

The proportions of the human body appear to change when the figure is not standing erect. As you begin studying figures in motion and from different viewpoints, you'll need to adjust the proportions to maintain a sense of realism.

FORM

Form refers to your subject's three-dimensional qualities, including depth and volume. Learning to represent form on paper is fundamental to any subject in classical drawing. If you don't master this, you risk the subject appearing flat, two-dimensional, and lacking life. The following pages feature studies that will train your eye to see simplified forms that make up the complex human body.

Movable Masses

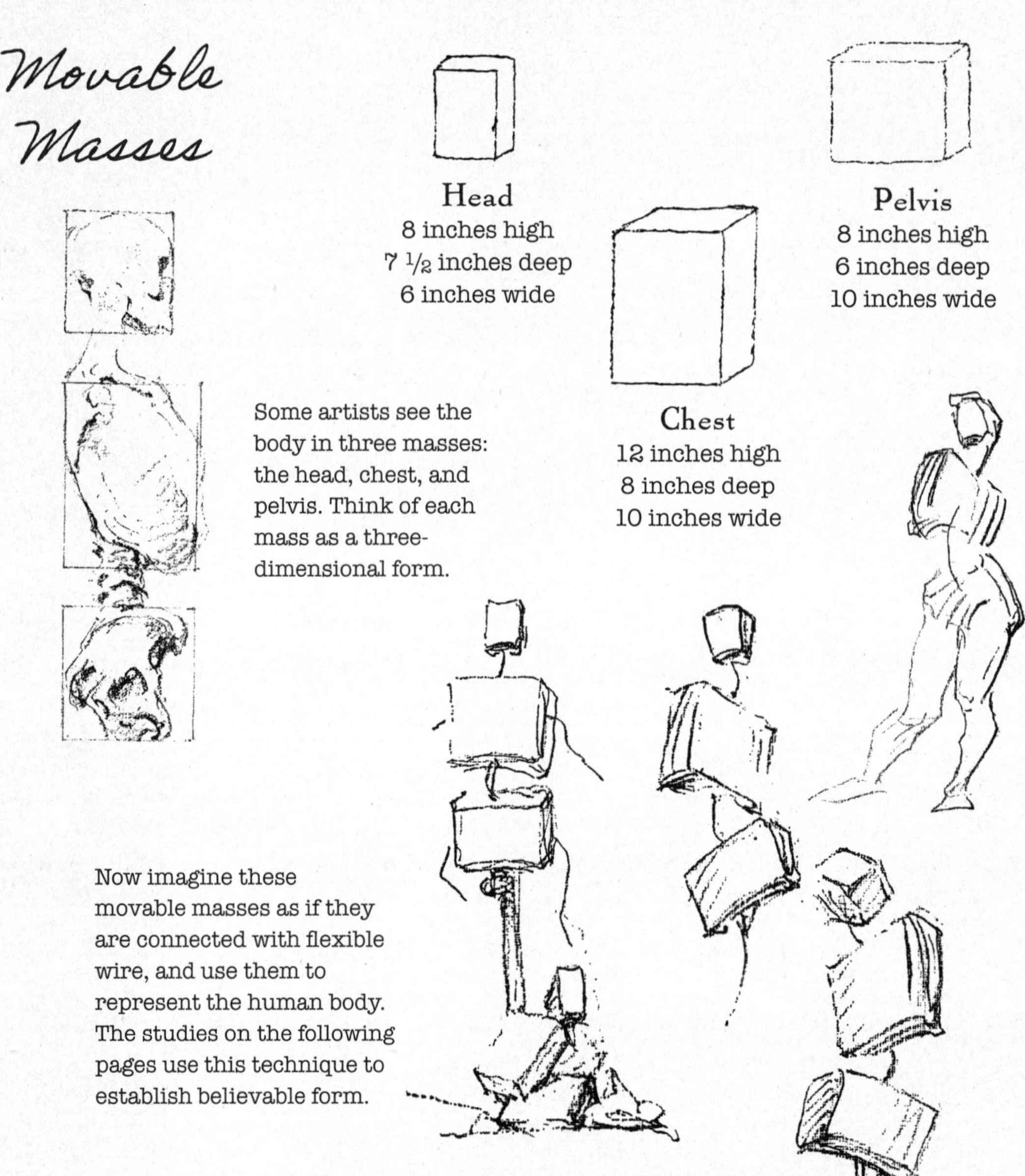

Head

8 inches high
7 ½ inches deep
6 inches wide

Pelvis

8 inches high
6 inches deep
10 inches wide

Chest

12 inches high
8 inches deep
10 inches wide

Some artists see the body in three masses: the head, chest, and pelvis. Think of each mass as a three-dimensional form.

Now imagine these movable masses as if they are connected with flexible wire, and use them to represent the human body. The studies on the following pages use this technique to establish believable form.

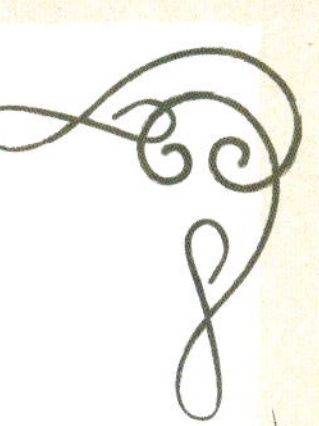

Use the practice pages in this section to draw a series of three-dimensional cubes for the head, chest, and pelvis in a variety of positions.

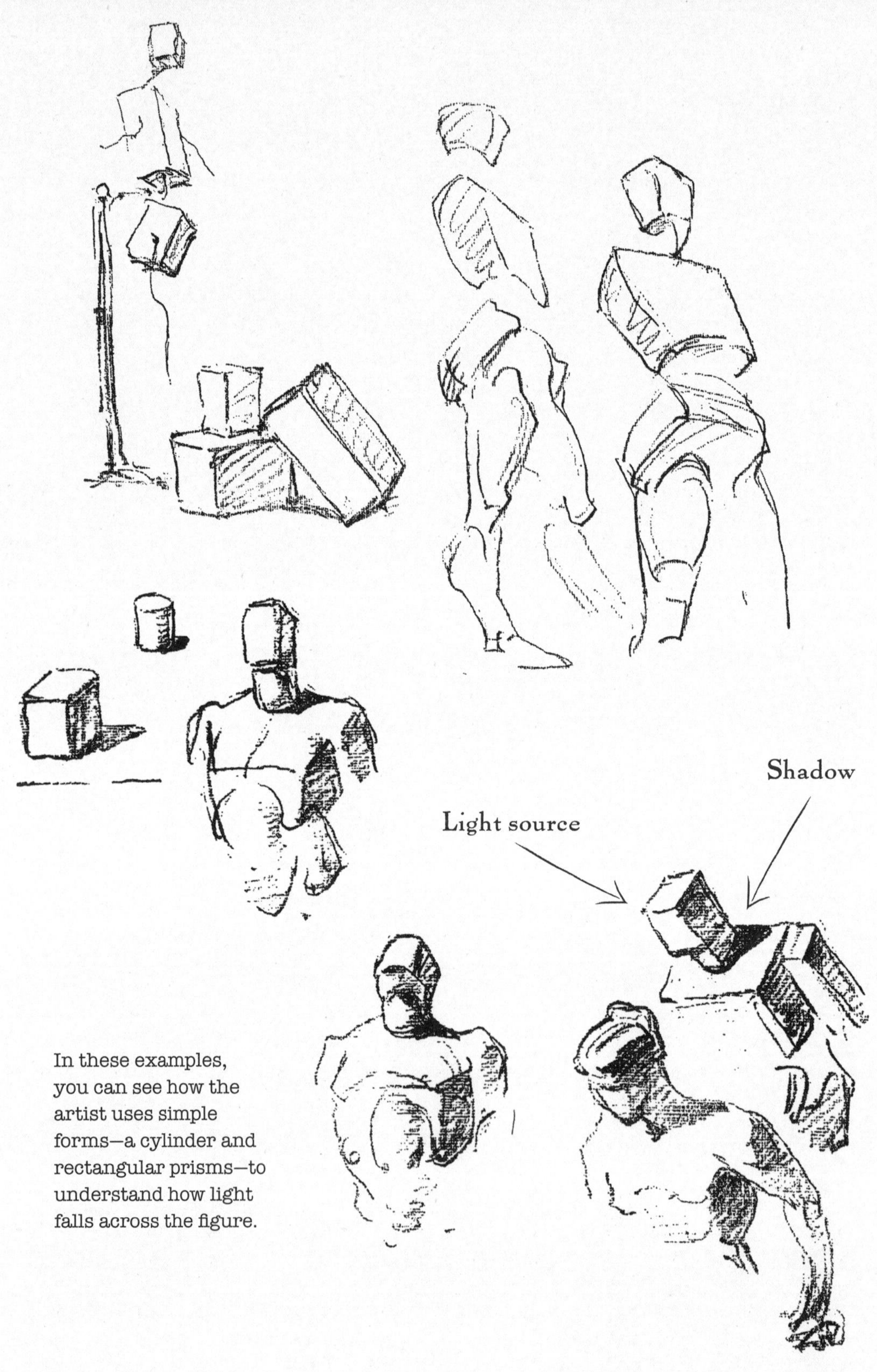

In these examples, you can see how the artist uses simple forms—a cylinder and rectangular prisms—to understand how light falls across the figure.

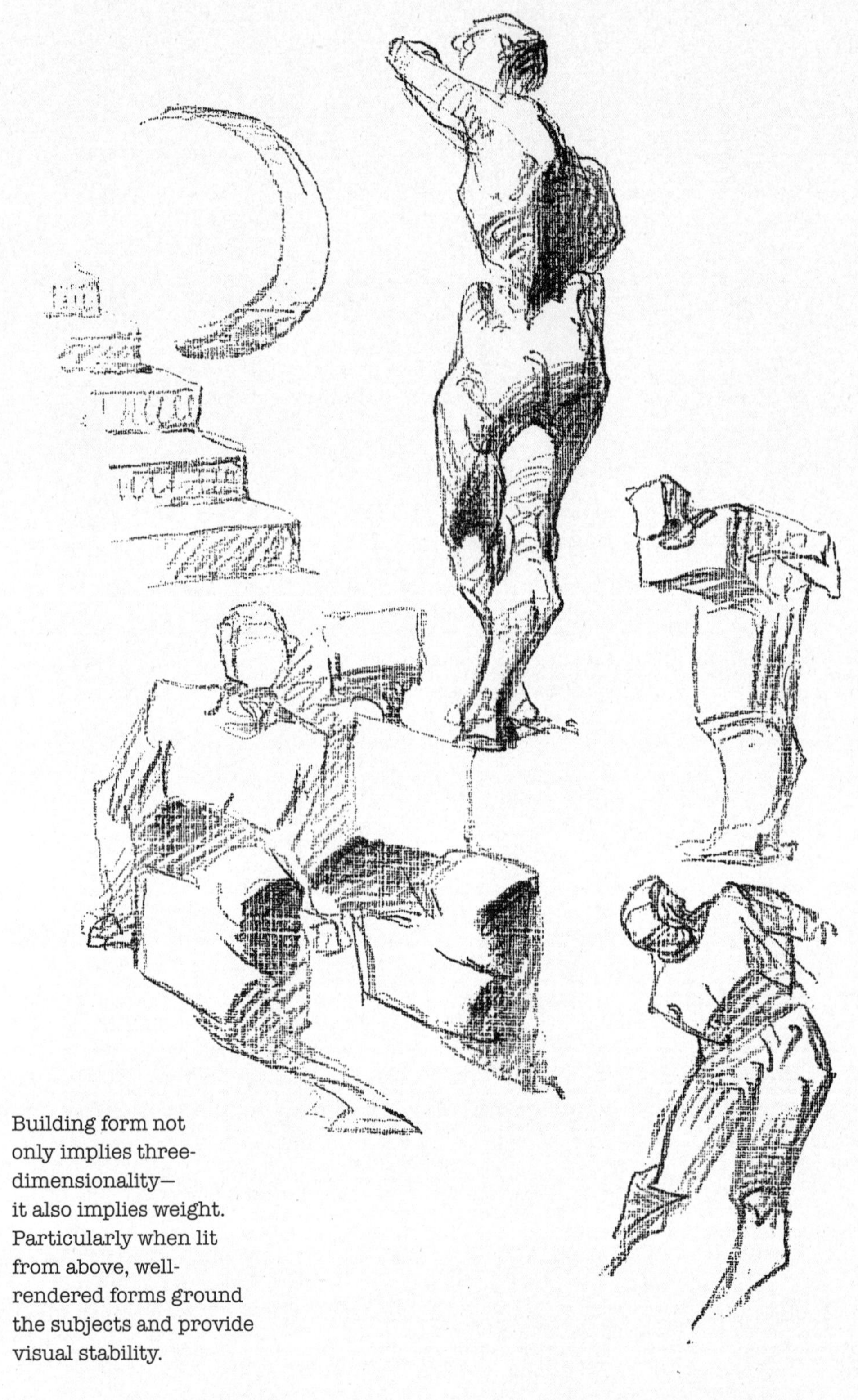

Building form not
only implies three-
dimensionality—
it also implies weight.
Particularly when lit
from above, well-
rendered forms ground
the subjects and provide
visual stability.

Moldings

Architectural moldings consist of curves, hollows, and ledges that add a decorative touch to buildings and furniture by means of light and shadow. Thinking of the human figure as molding can be useful in understanding how light and shadow define the undulating, varied forms of the body.

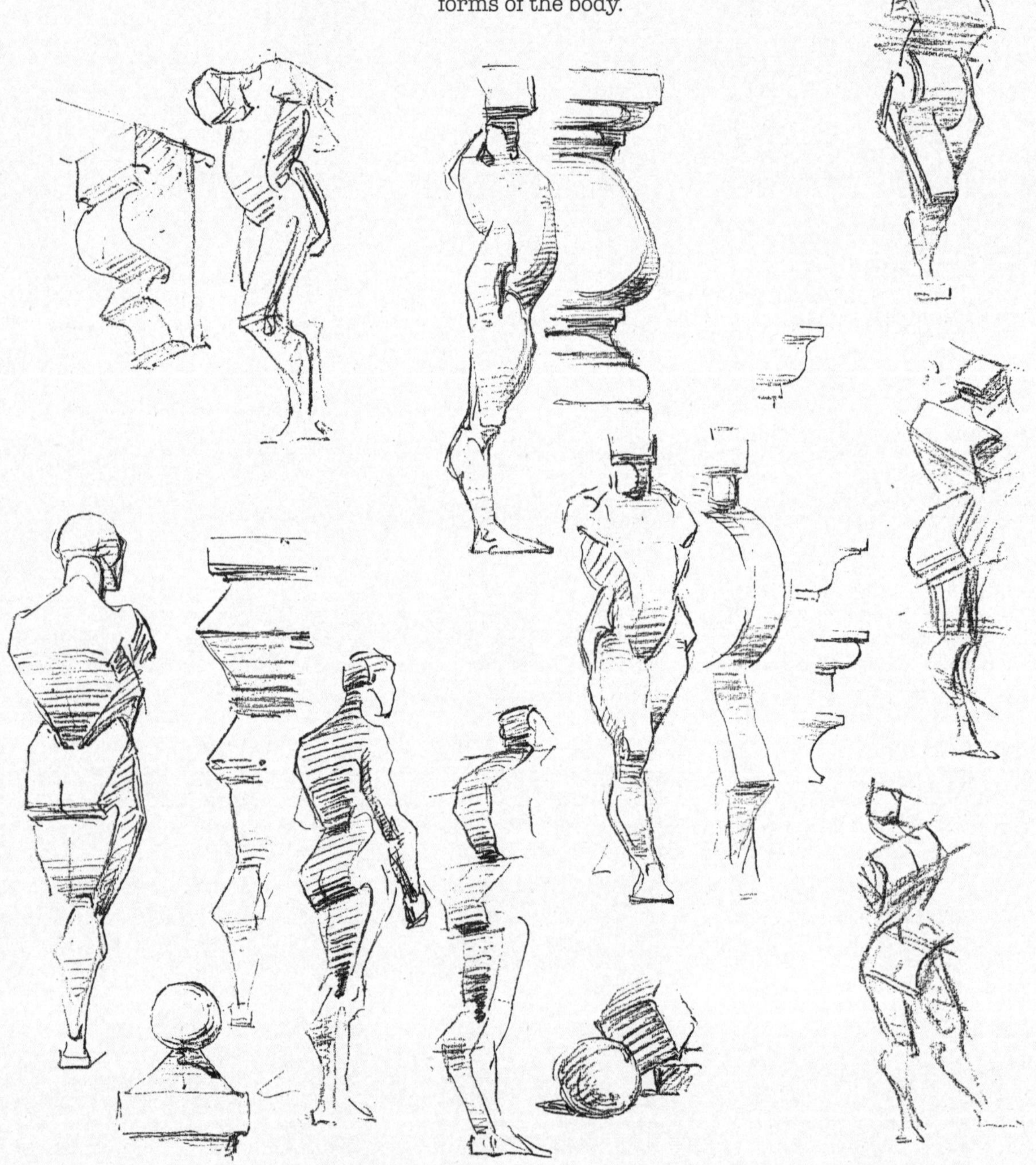

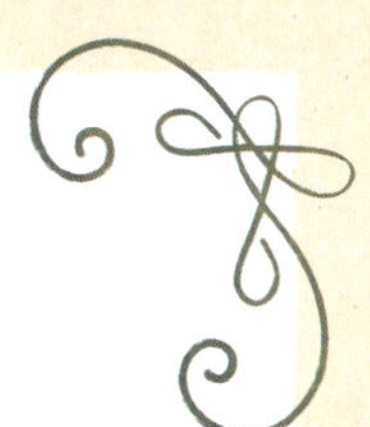

Sketch some architectural moldings, noting the areas of light and shadow. How can you use these studies to help you draw realistic figures?

BALANCE

When several objects are balanced at different angles, one above the other, they have a common center of gravity. In a drawing, there must be a sense of security—balance between the opposite or counteracting forces—regardless of where the centerline may fall. Great practice for establishing realistic poses lies in the balance of the figure.

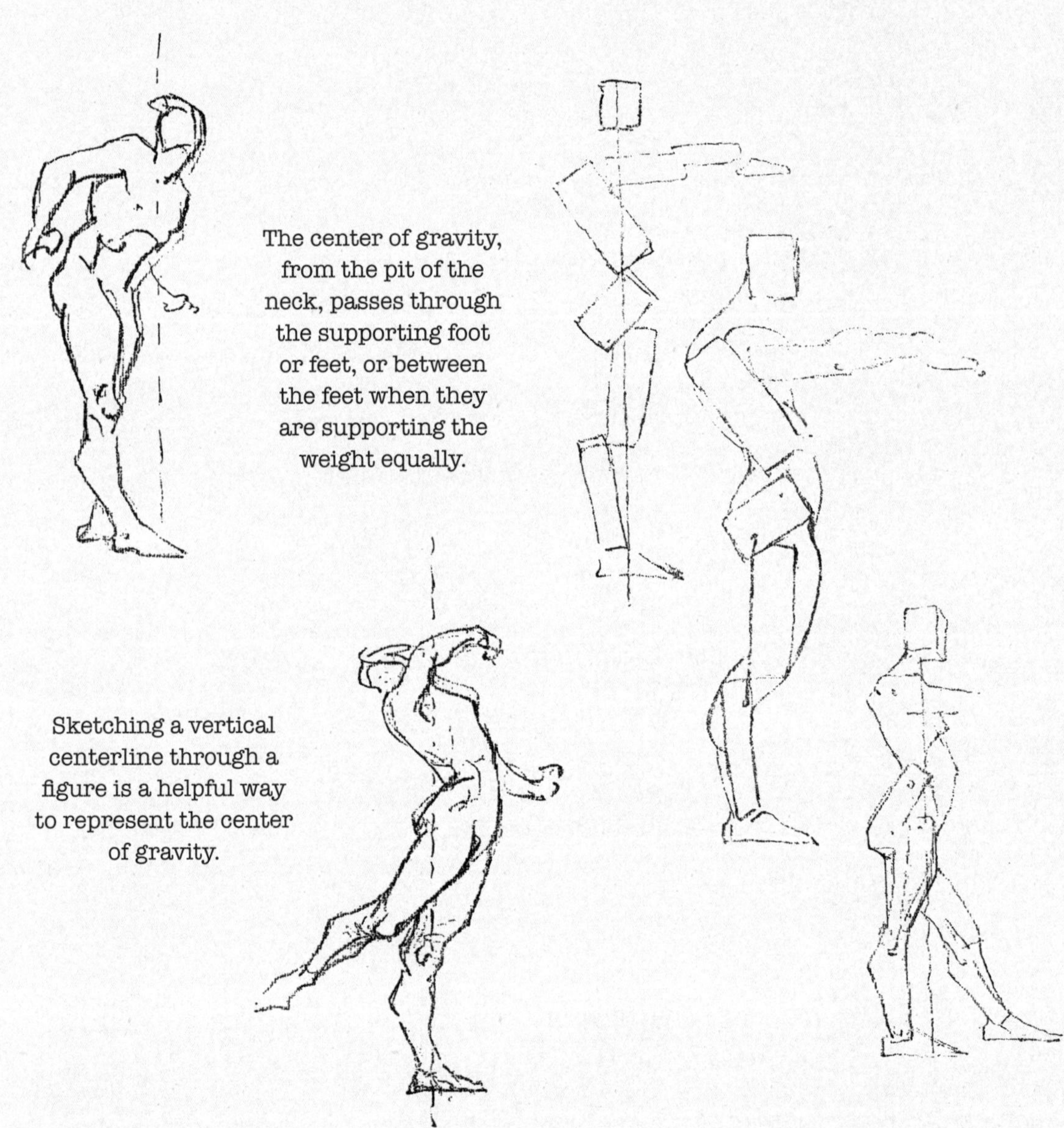

The center of gravity, from the pit of the neck, passes through the supporting foot or feet, or between the feet when they are supporting the weight equally.

Sketching a vertical centerline through a figure is a helpful way to represent the center of gravity.

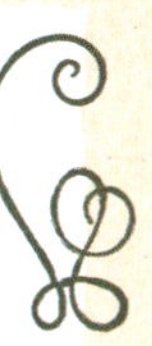

Use the practice pages in this section to create gestural drawings with centerlines. How can you improve your technique so that each pose looks balanced?

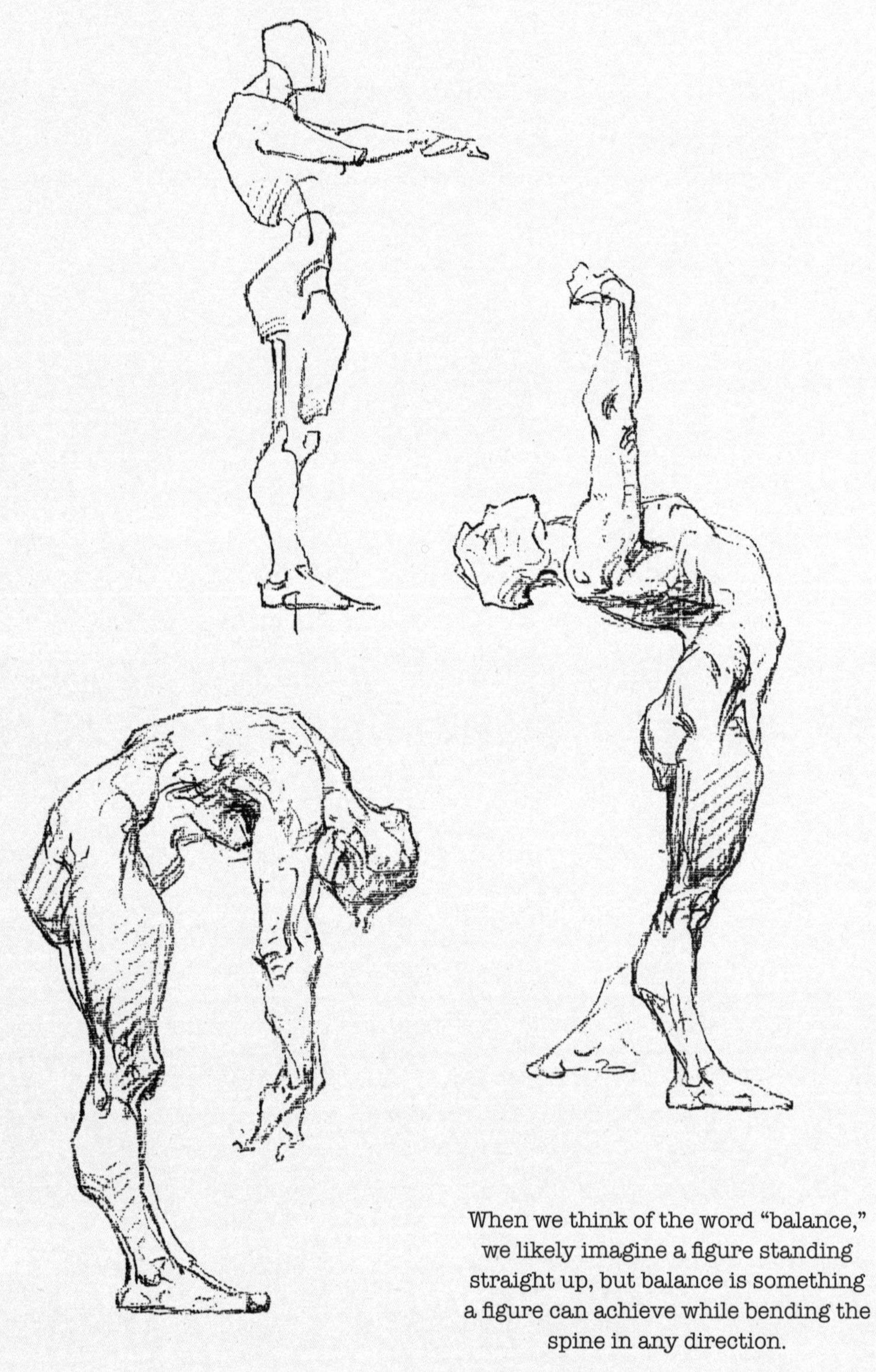

When we think of the word "balance," we likely imagine a figure standing straight up, but balance is something a figure can achieve while bending the spine in any direction.

Finding the Balance

One of the best ways to study balance and poses
is to watch dancers, gymnasts, and acrobats. Due
to their difficult movements, they must maintain
control of their bodies, constantly finding their
center of balance.

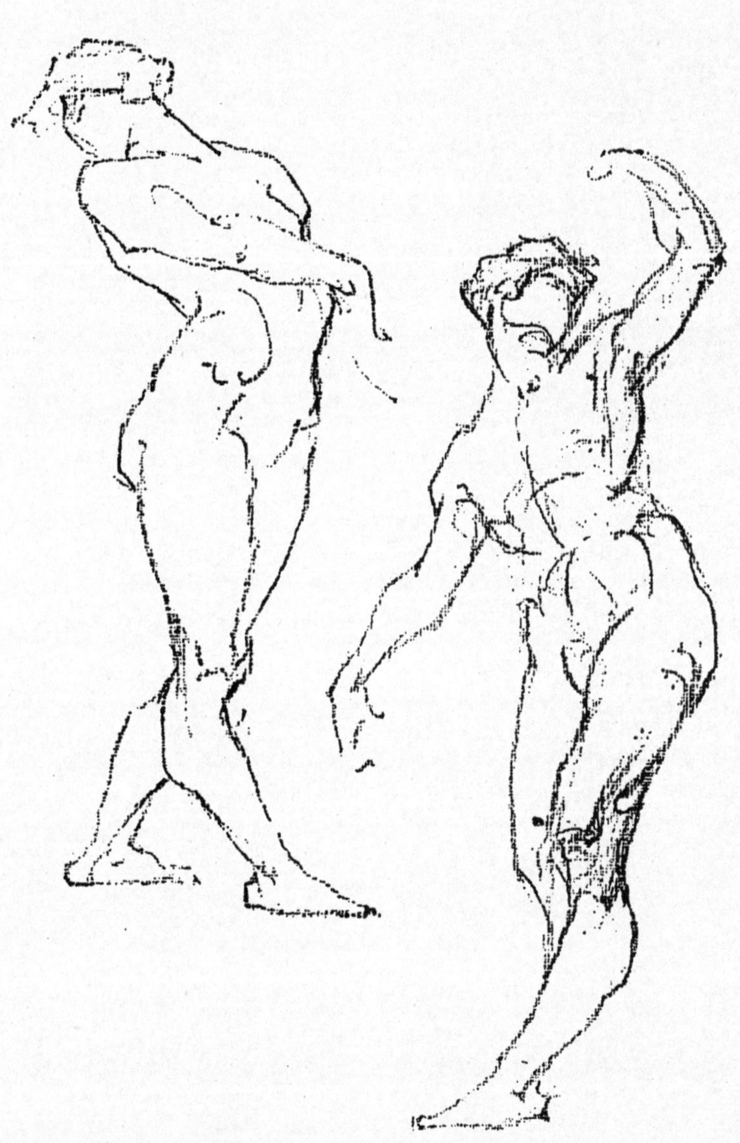

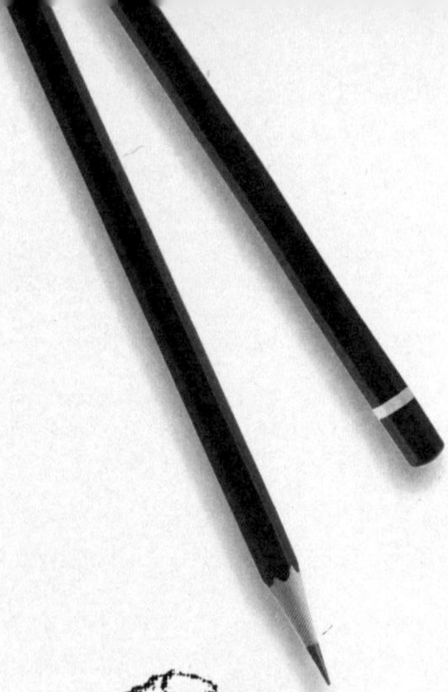

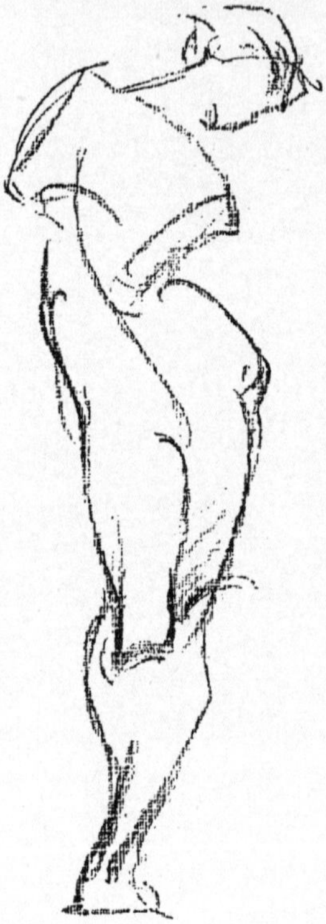

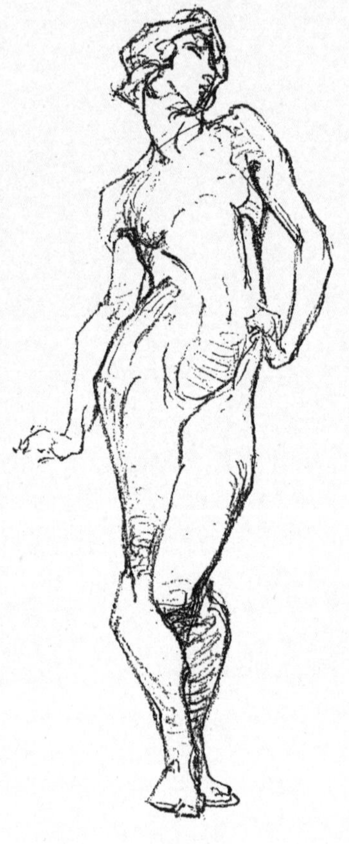

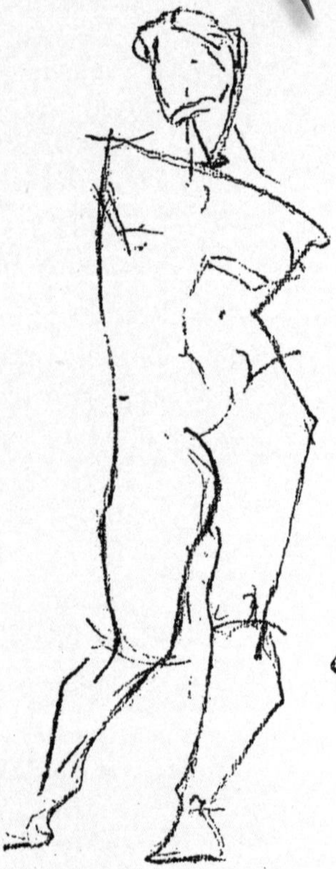

Contrapposto

Contrapposto is an Italian term that means "counterpose." The contrapposto stance causes a twist of the hips and a curve of the spine, giving the body an organic *S* shape. First mastered by the Greeks, this relaxed, balanced pose is celebrated throughout classical drawing, painting, and sculpture, such as in Michelangelo's *David*.

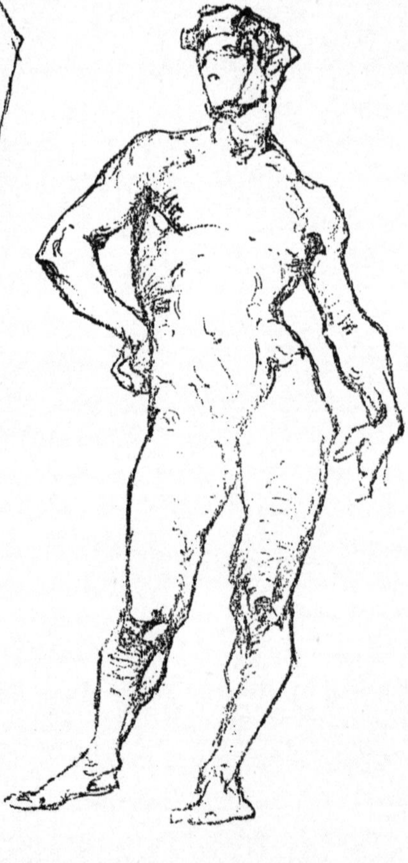

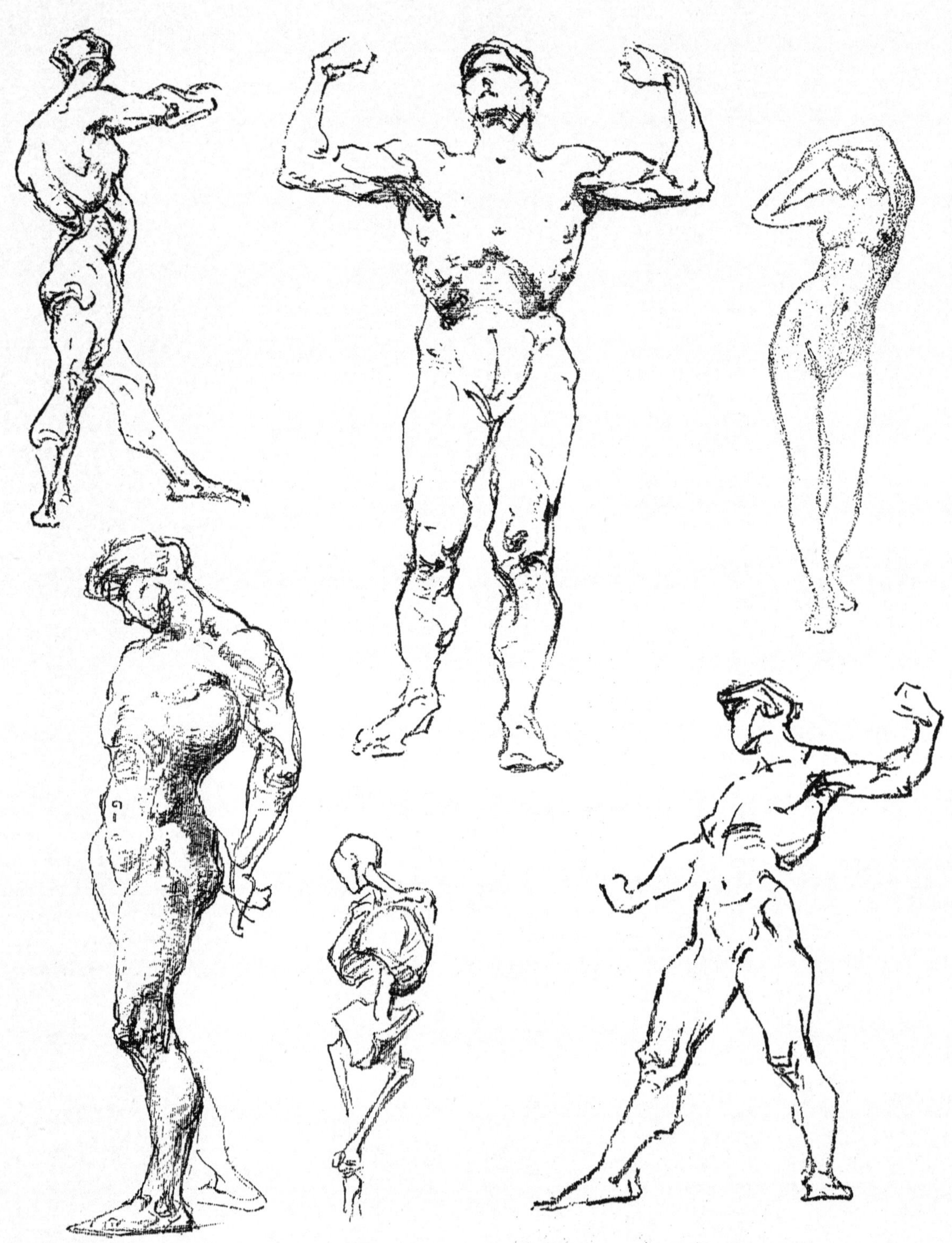

MOVEMENT

Composition refers to the placement of elements within the frame of your drawing. Good compositions create a pleasing pattern and lead the viewer's eye around the scene. It's a great idea to create small sketches, called "thumbnails," to work out your basic composition before beginning a drawing.

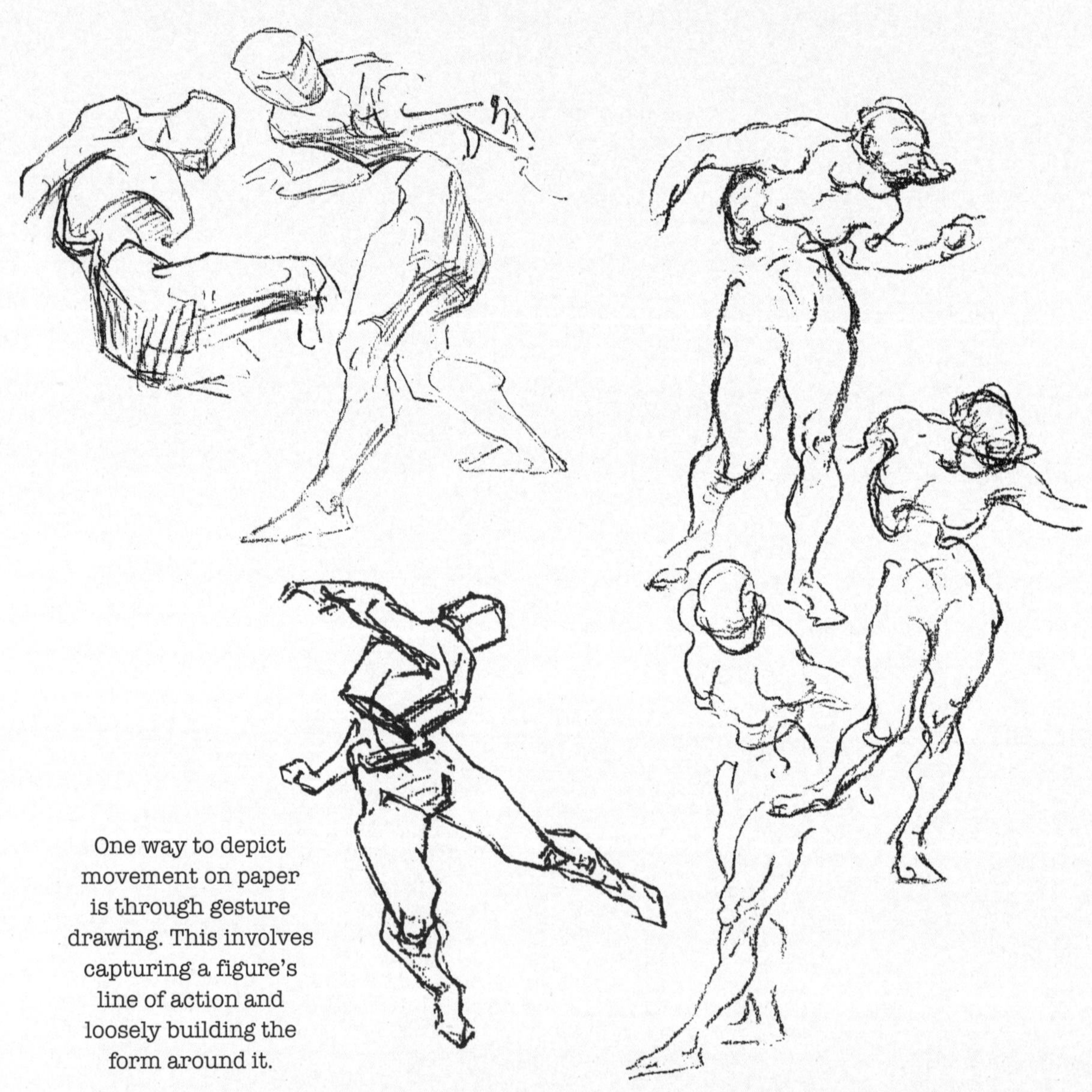

One way to depict movement on paper is through gesture drawing. This involves capturing a figure's line of action and loosely building the form around it.

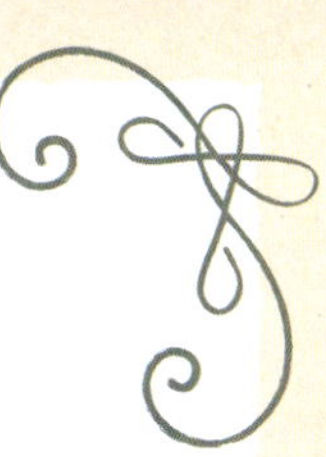

Use the practice pages in this
section to create a series of
thumbnail drawings of figures
in various states of motion.

Rhythm

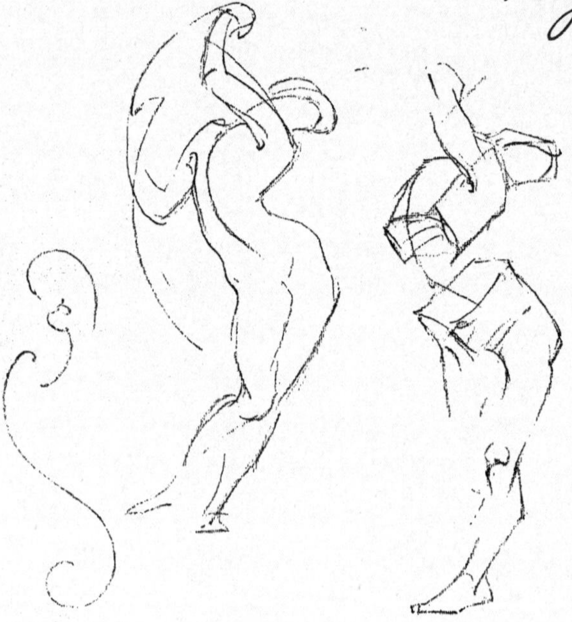

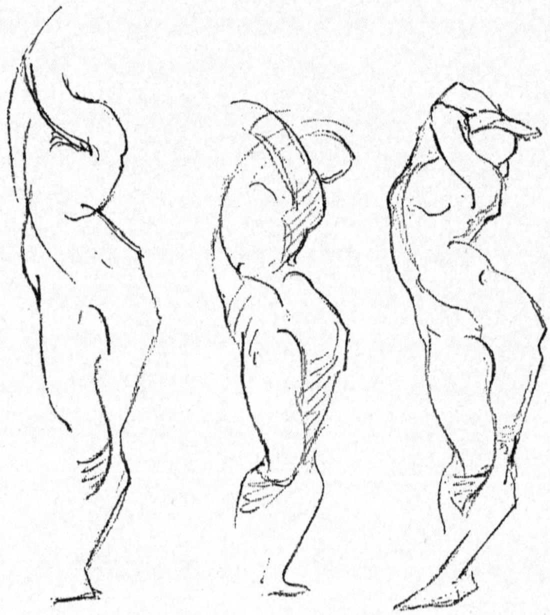

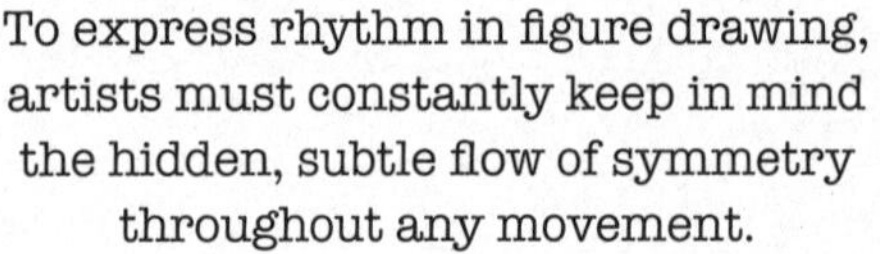

To express rhythm in figure drawing,
artists must constantly keep in mind
the hidden, subtle flow of symmetry
throughout any movement.

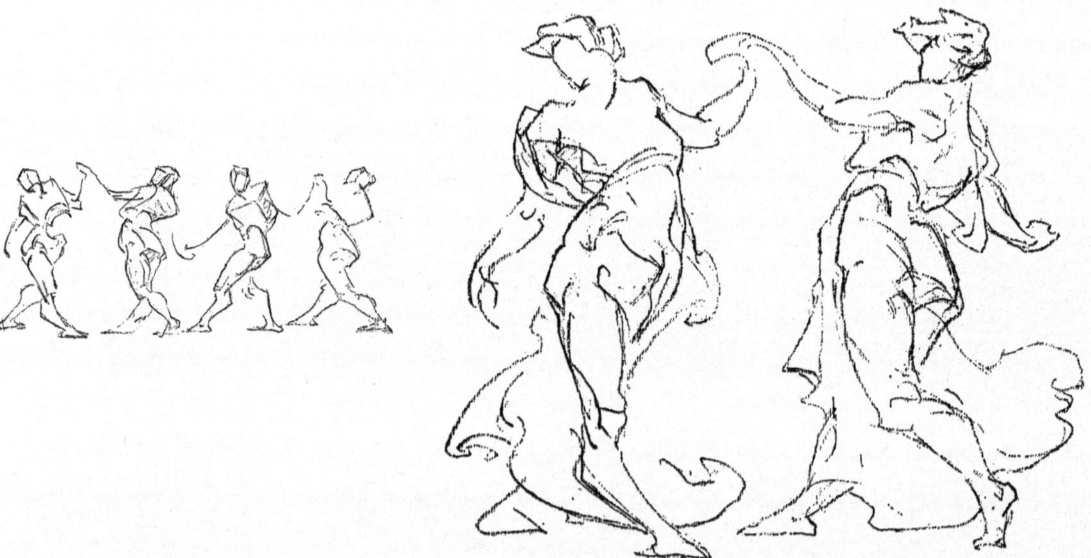

Twisting & Turning

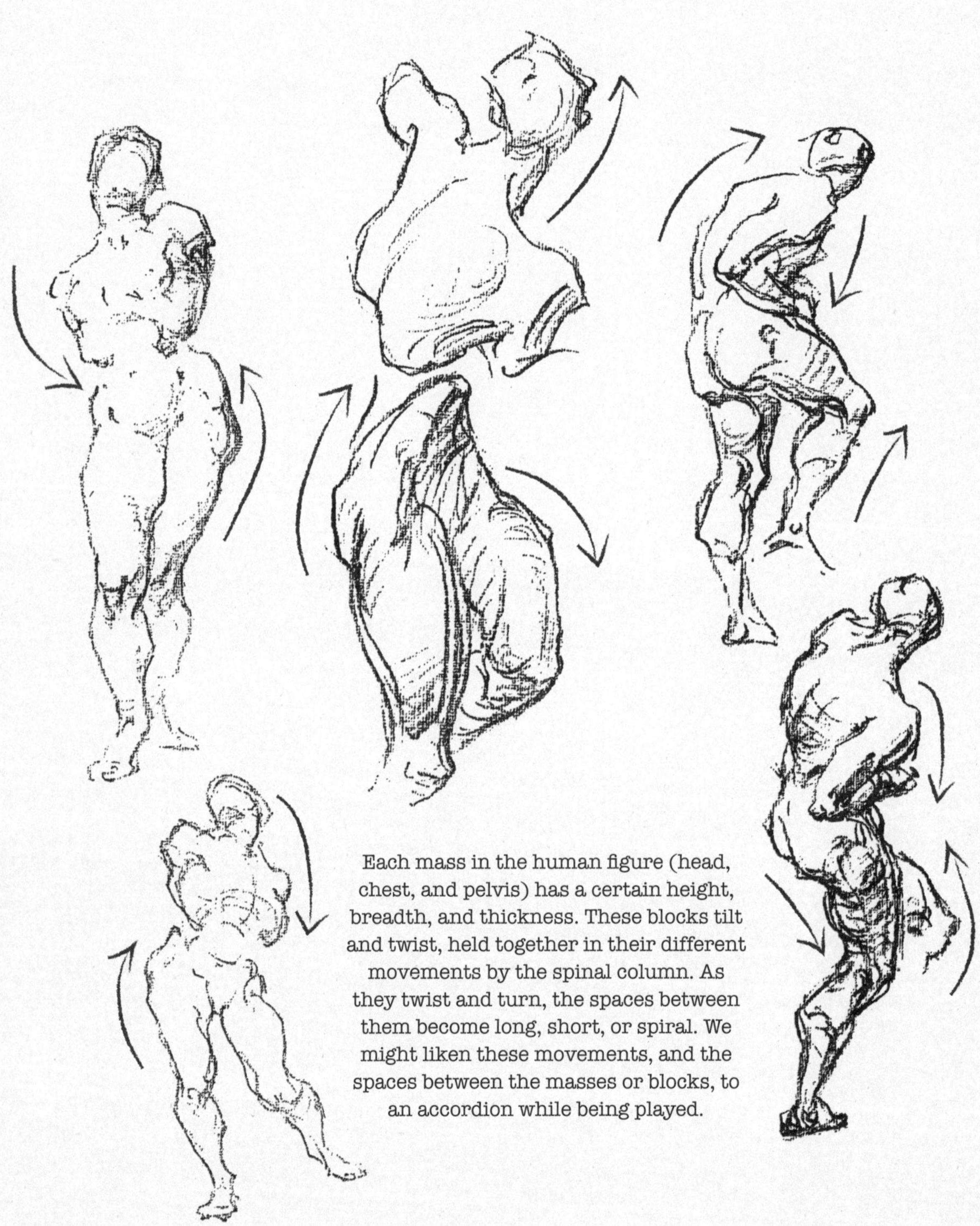

Each mass in the human figure (head, chest, and pelvis) has a certain height, breadth, and thickness. These blocks tilt and twist, held together in their different movements by the spinal column. As they twist and turn, the spaces between them become long, short, or spiral. We might liken these movements, and the spaces between the masses or blocks, to an accordion while being played.

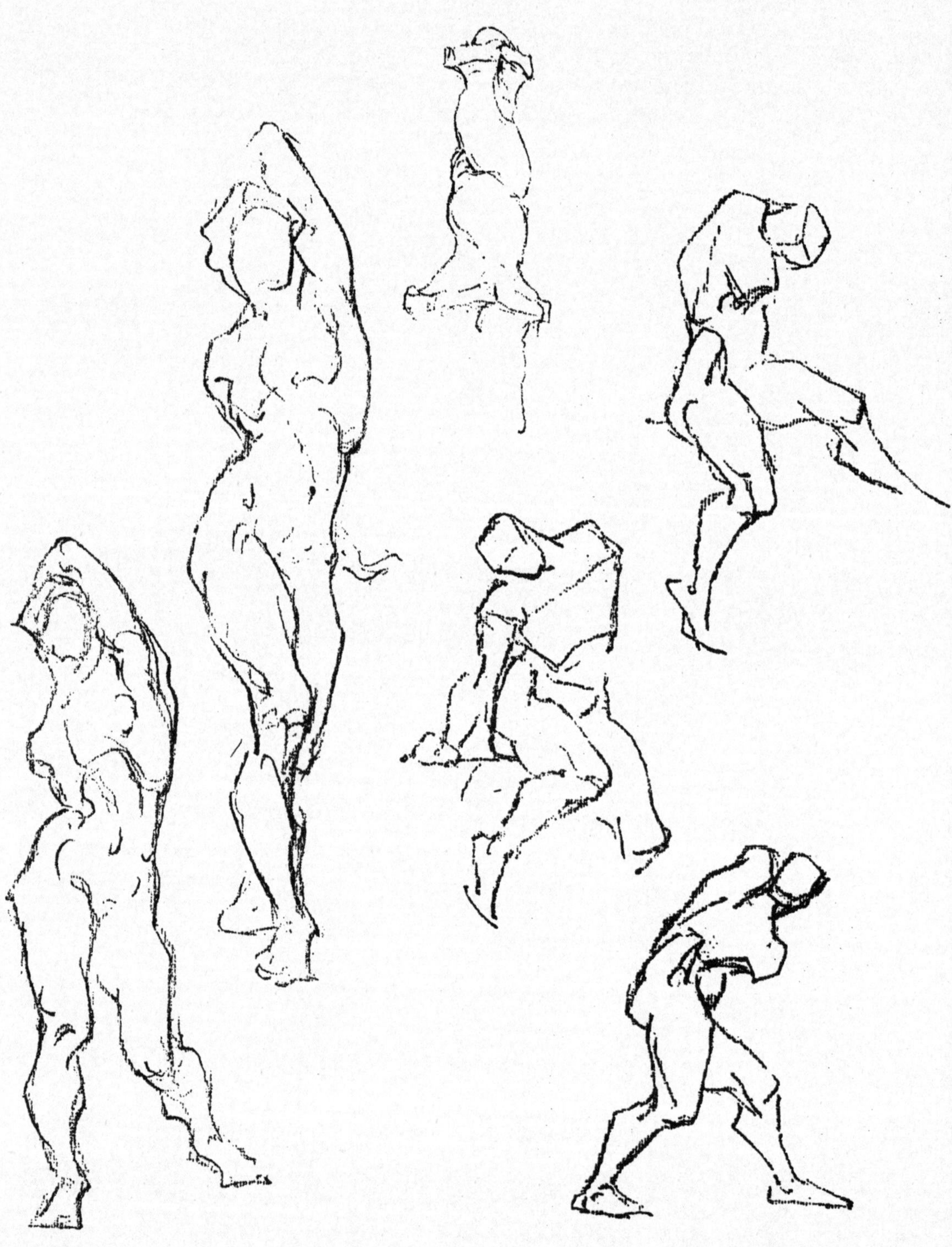

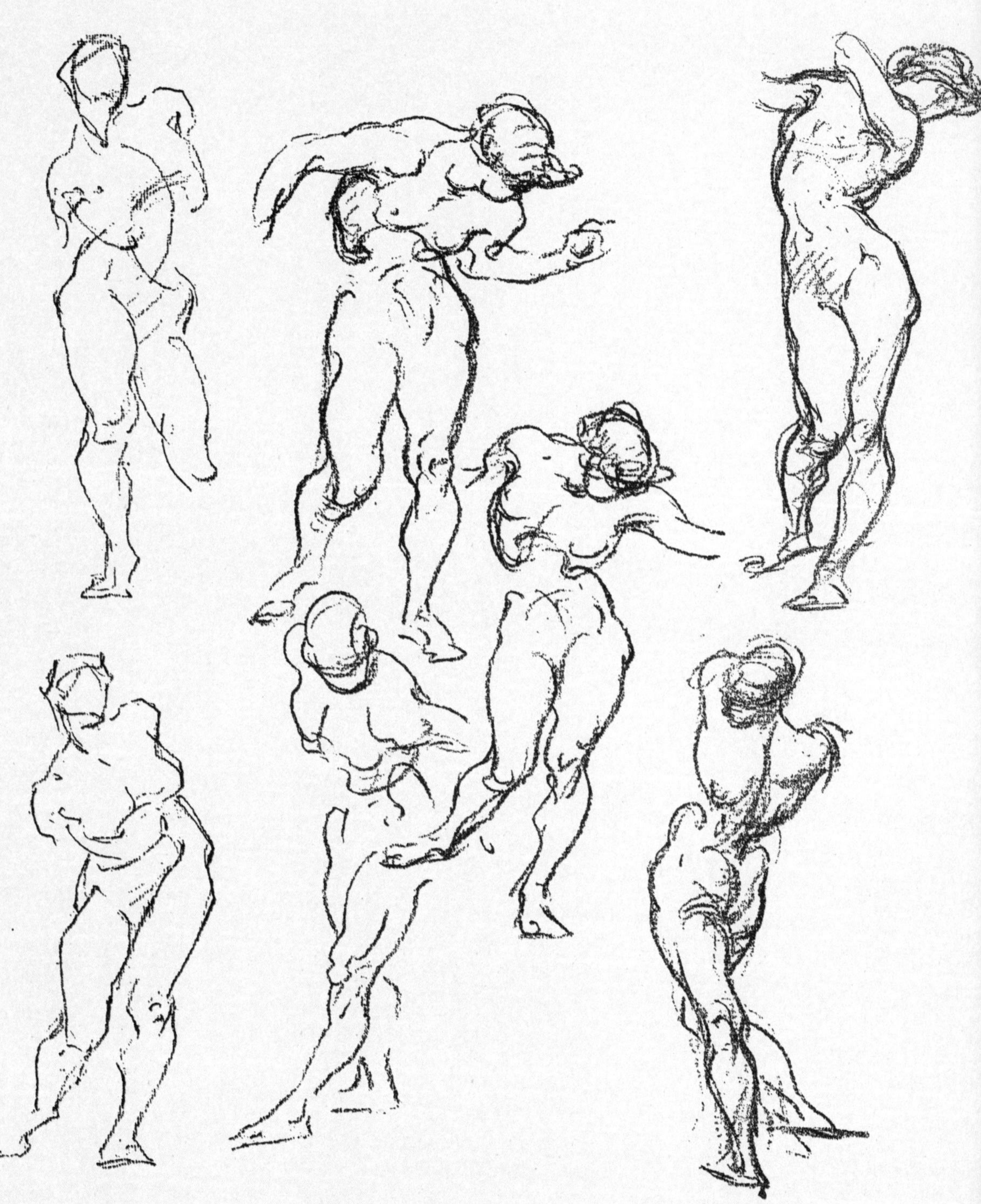

Seeing Form with Mechanics

To help you better represent accurate movement, it can be helpful to envision the moving body part as a tool. Below is the hand showing how specific movements might relate to the mechanical action of tools.

Hook   Tongs   Scoop

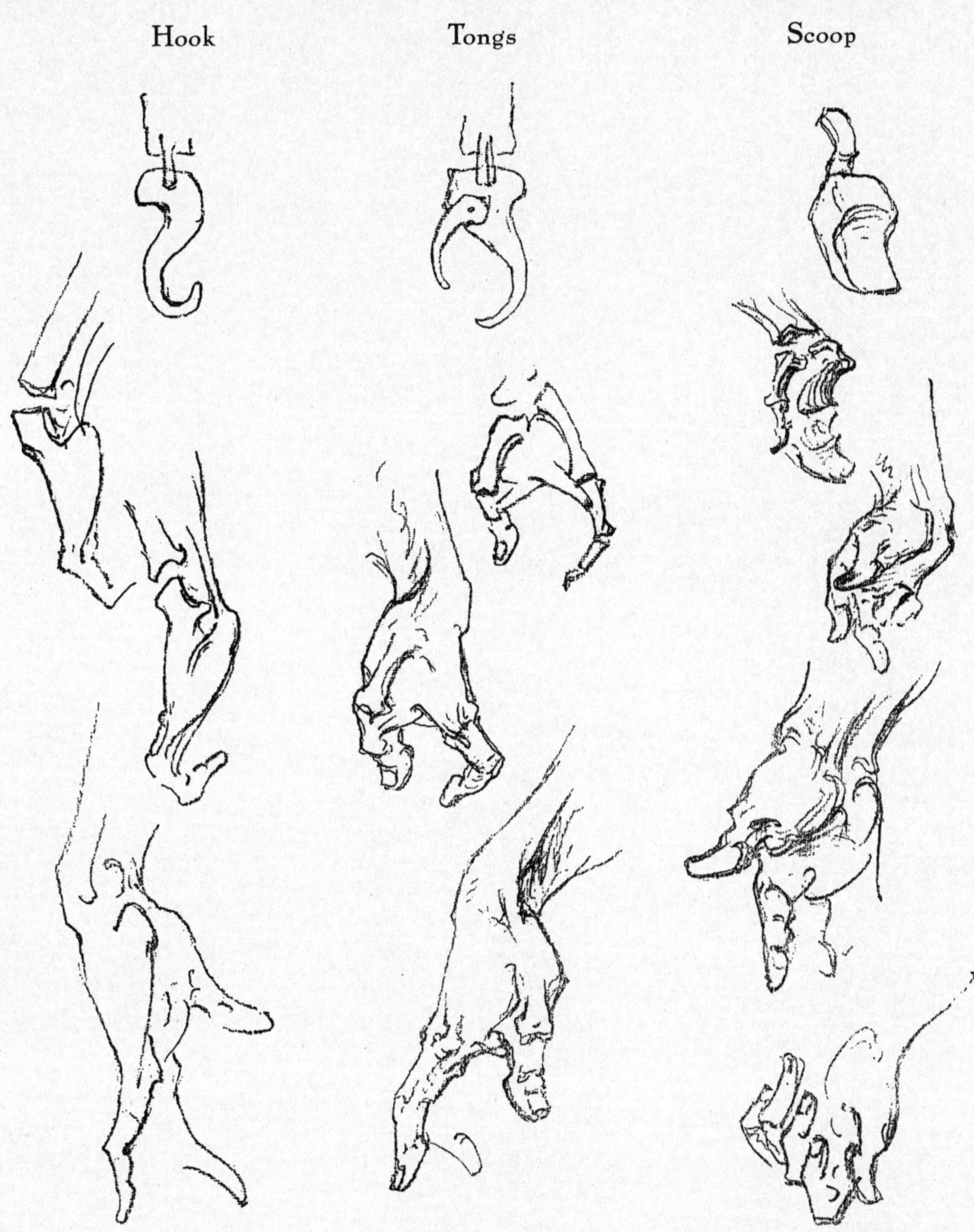

What other tools or machine parts move in ways similar to humans? Draw them here.

THE TORSO

The torso—also called the "trunk"—may be divided into three main masses: 1) the shoulders and chest, 2) the small of the back and abdominal region, and 3) the mass of the hips and pelvis. The first and third masses are defined primarily by bony structures, whereas the second mass is defined primarily by musculature and flesh.

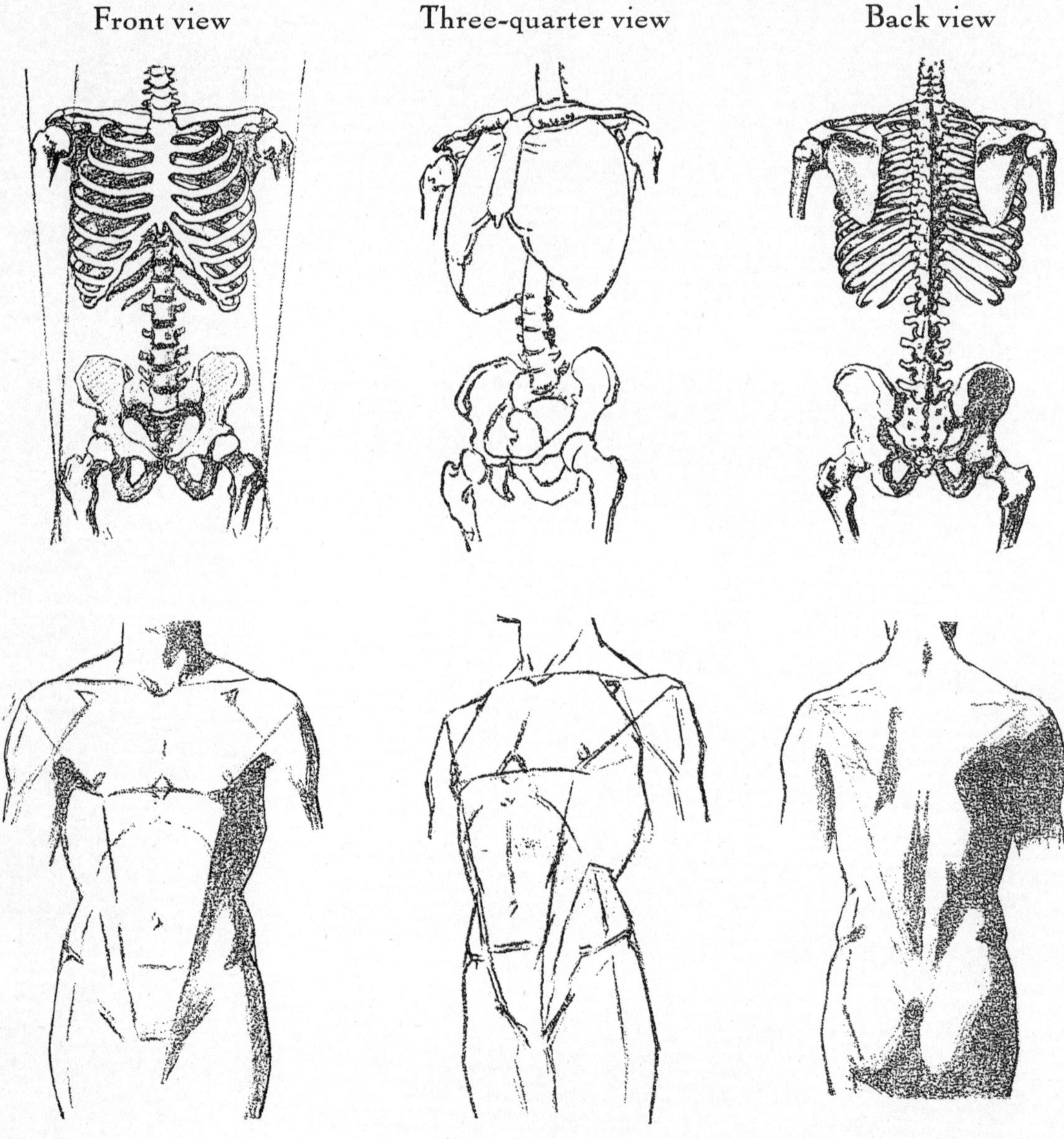

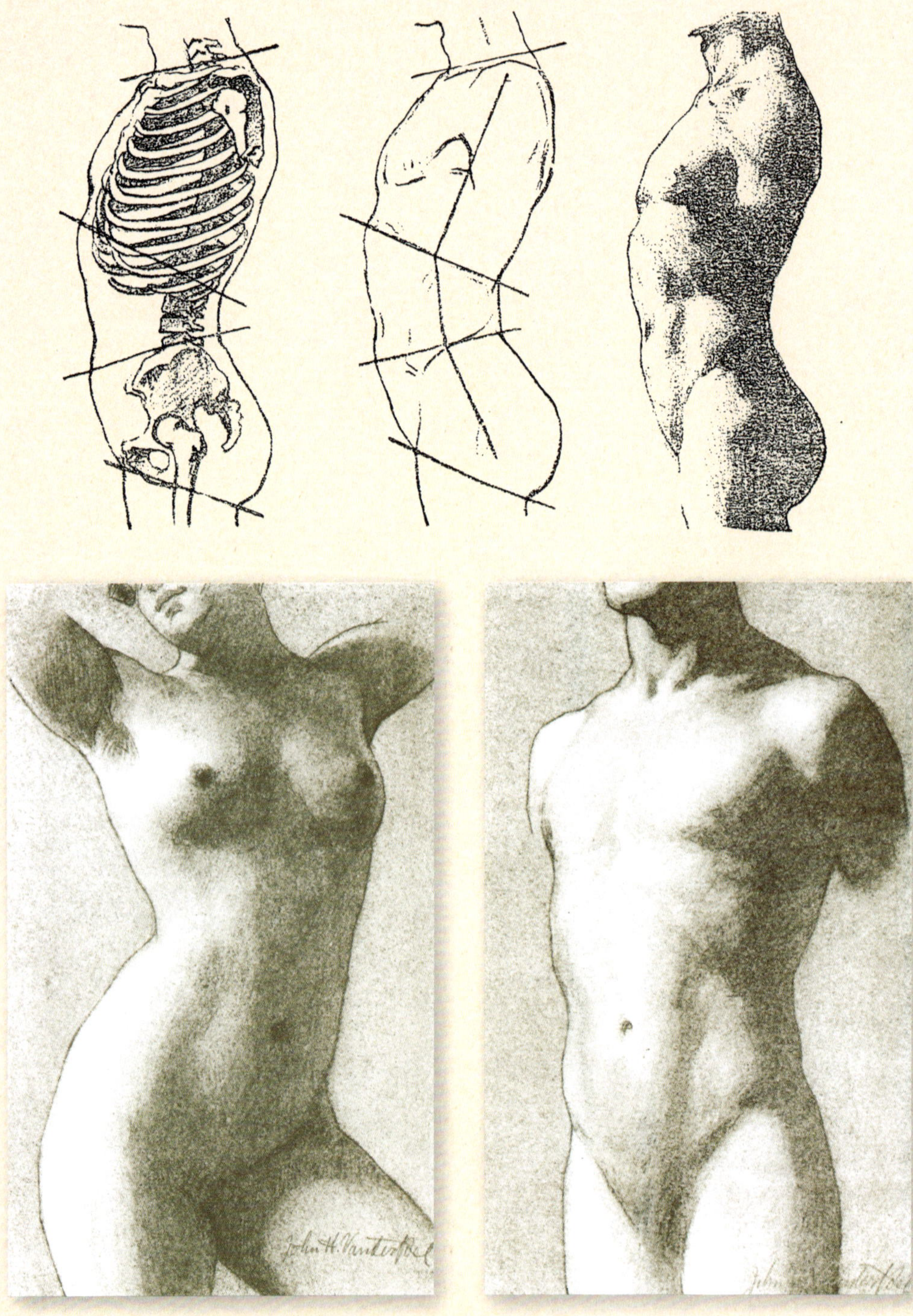

Dramatic side lighting highlights the differences between
typical male and female bodies.

The Male Trunk

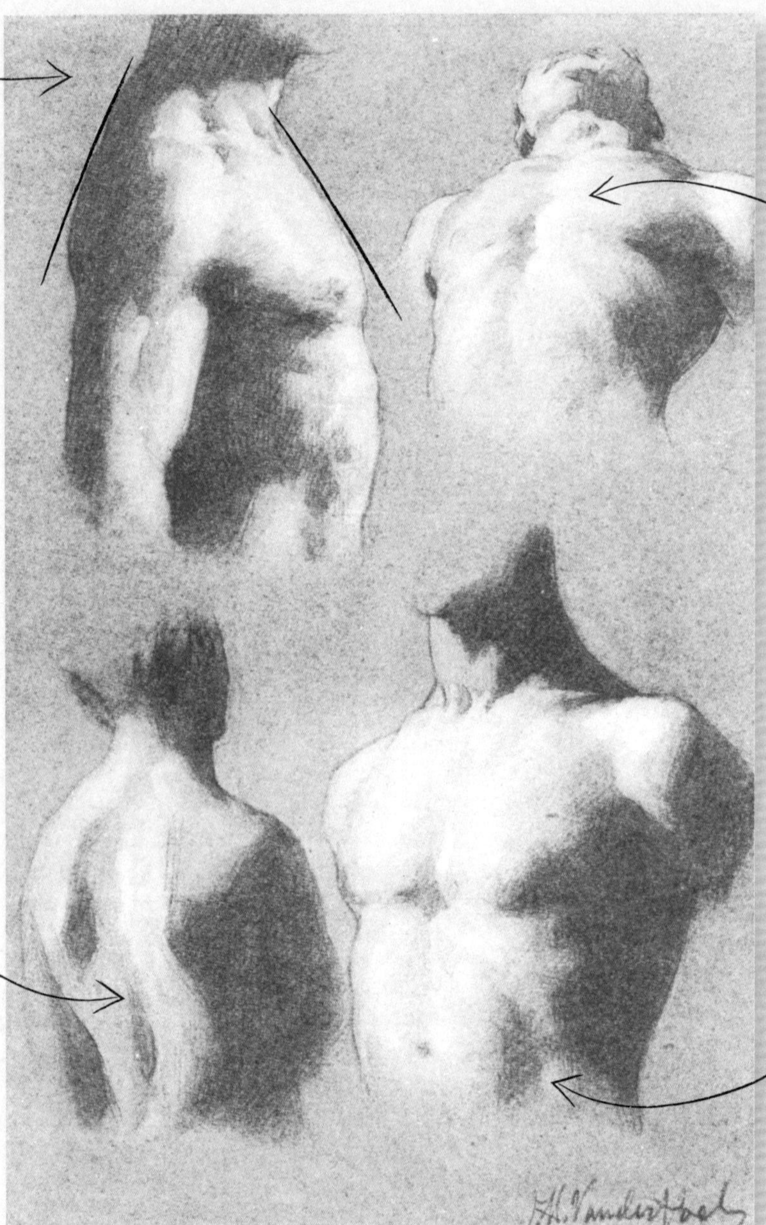

Profile
From a side view, the curvature of the spine is often obscured by the arm. Note the angles of the upper back and chest originating from the base of the neck.

The Back
A three-quarter view from behind shows the *S*-shaped length of the spine as well as the rounded points of the scapula.

The Chest
The cage-like form of the ribs becomes more square as it descends, and this angularity is greatly increased by the cushion-like form of the pectoral muscles in front.

The Abdomen
In this example, the dramatic side lighting defines the abdominal musculature below the rib cage

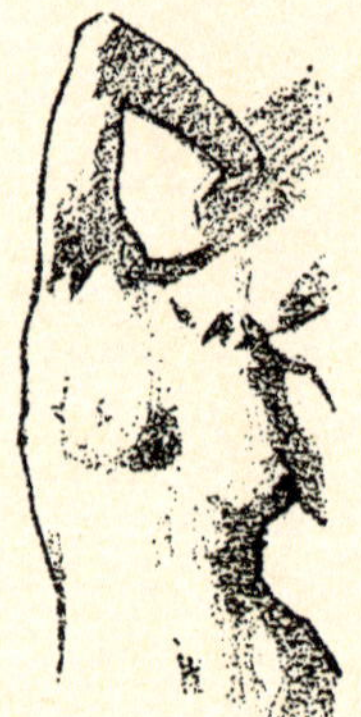

The female rib cage is (on average) 10 percent smaller than the male rib cage. However, female breasts can add to the mass of the chest and give the appearance of a narrower waist.

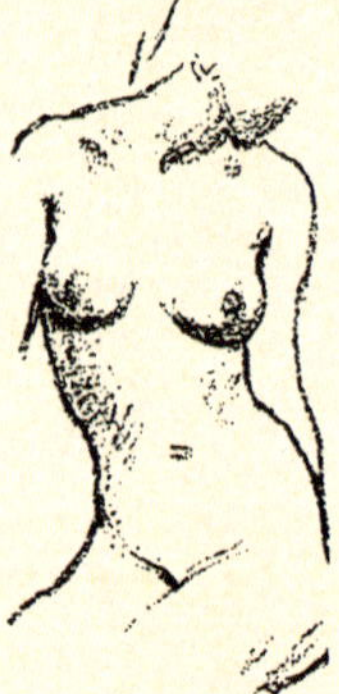

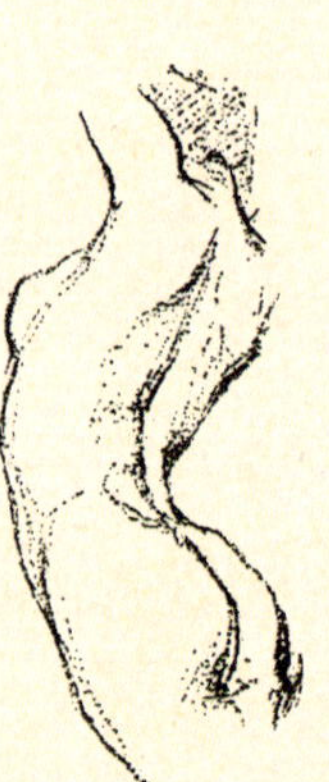

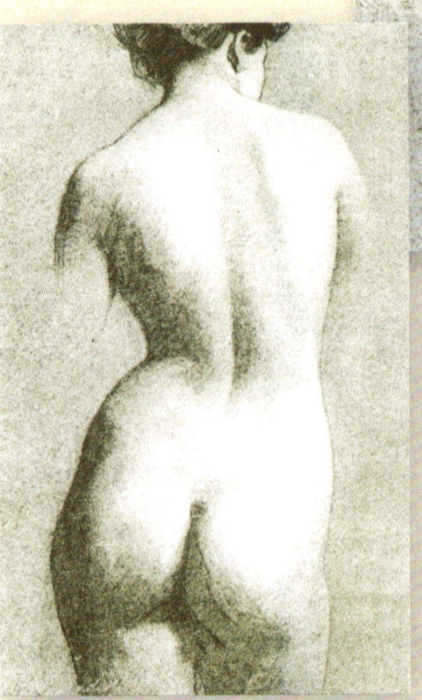

The pelvic mass of a female body is typically larger in relation to the rest of the body. This gives most women a higher waistline.

More Studies of the Torso

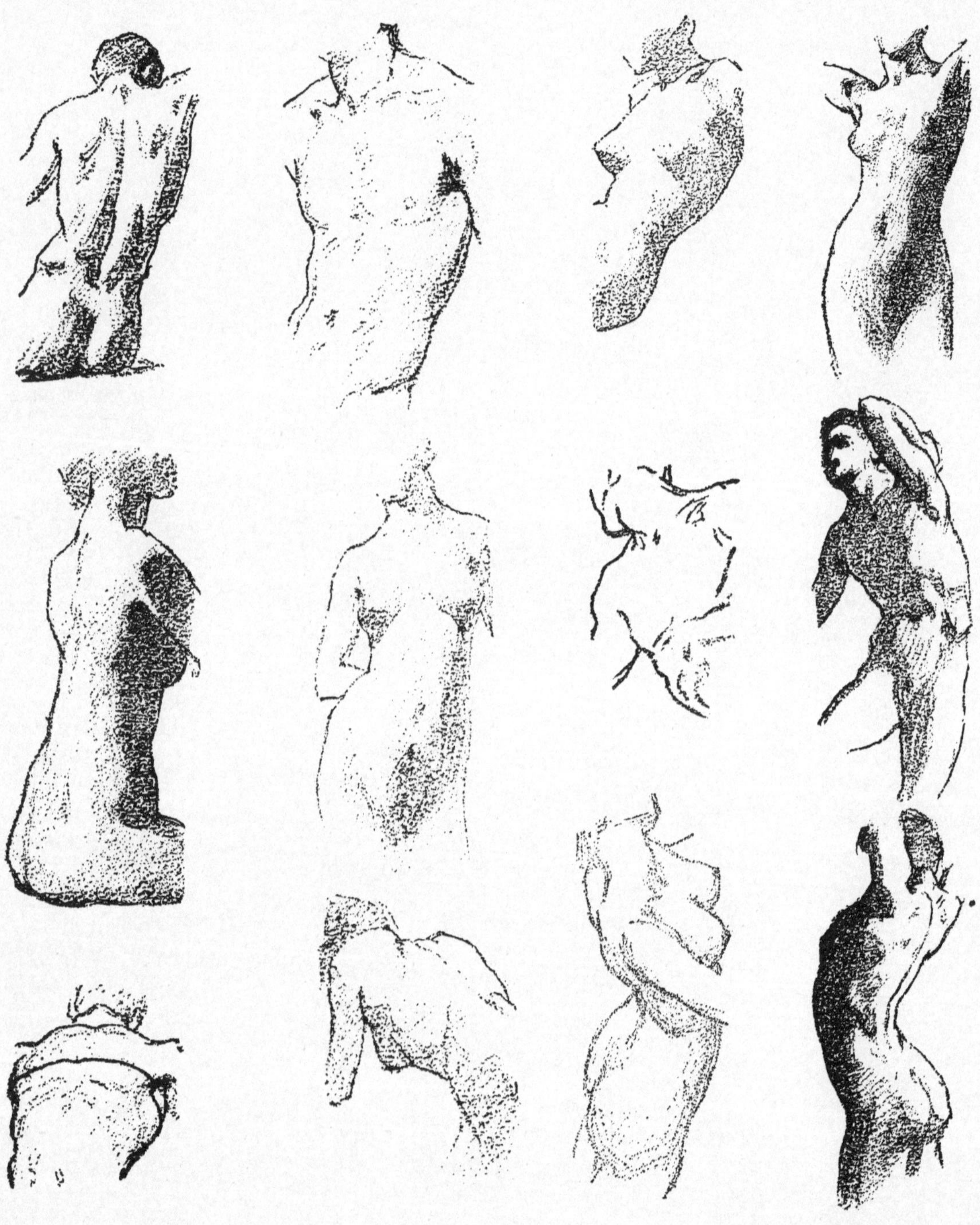

ARMS & HANDS

Think of arms and hands as supporting characters to a figure's emotional expression. Unlike the face, the arms give expressive clues that can be seen at a great distance. They can communicate everything from languid relaxation to joy and distress.

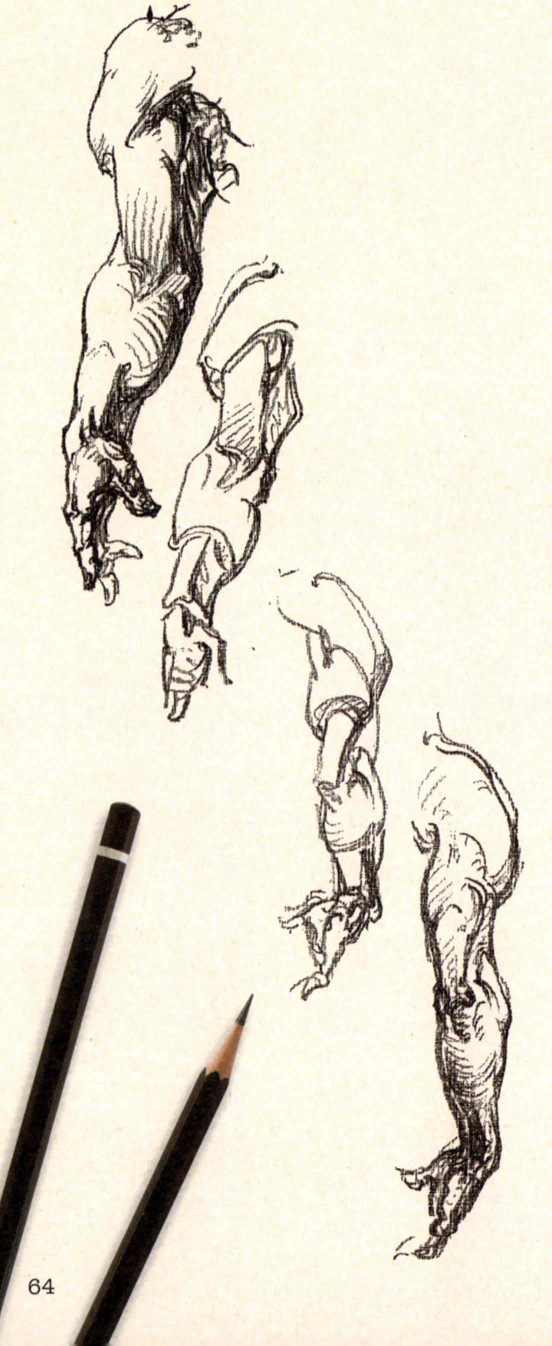

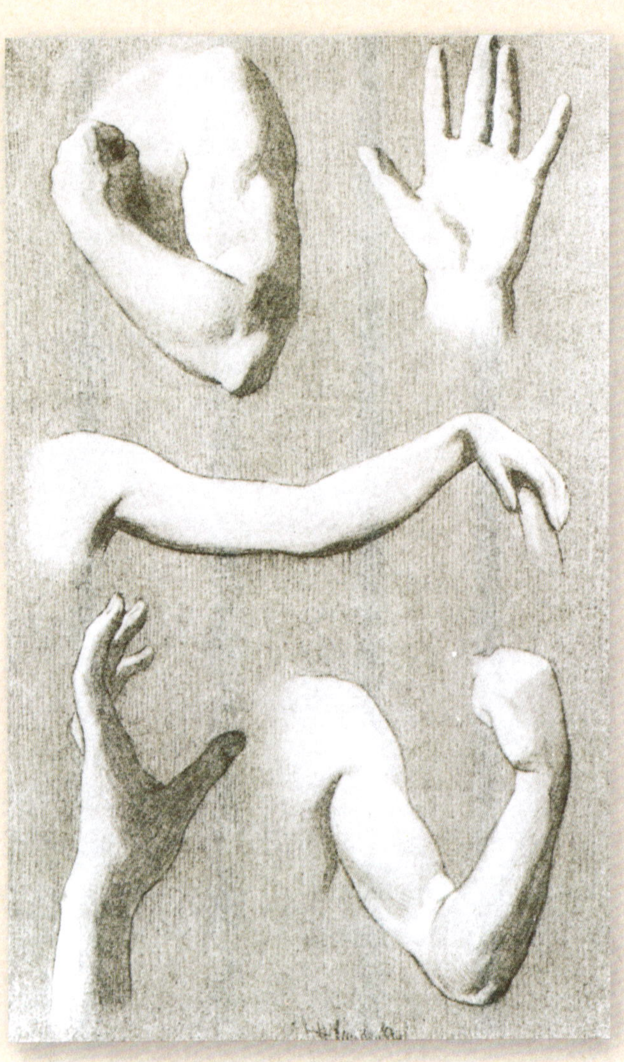

Male arms generally present more muscle mass and more defined musculature. Women generally have finer bones with softer contour lines and slender, tapering fingers.

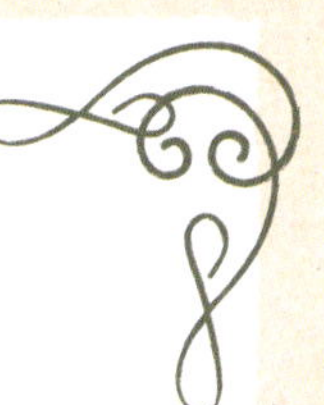

Hands are one of the more challenging things to master in figure drawing. Use the practice pages in this section to draw hands and arms in different poses.

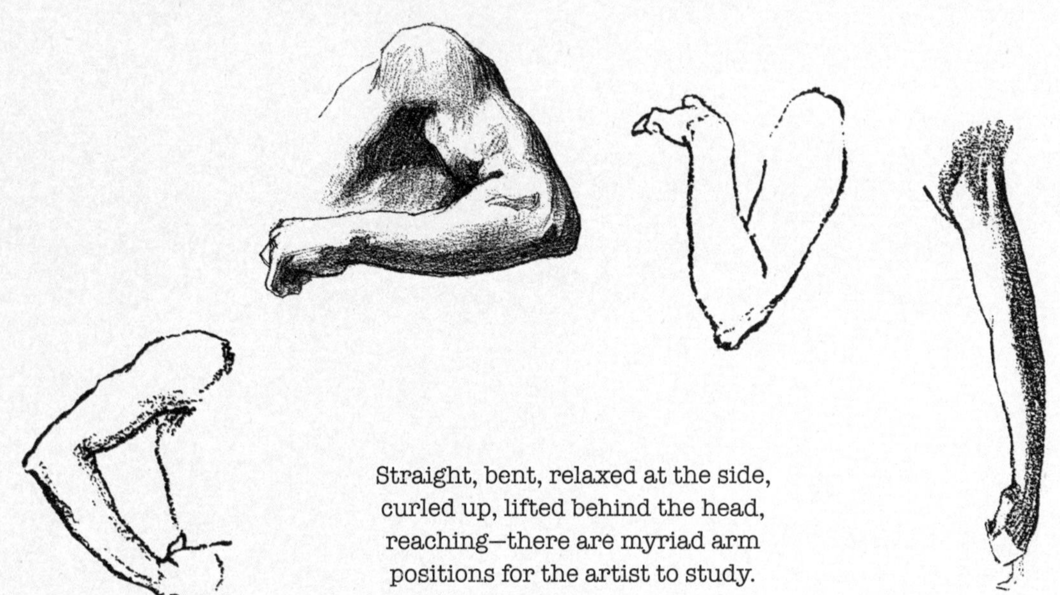

Straight, bent, relaxed at the side,
curled up, lifted behind the head,
reaching—there are myriad arm
positions for the artist to study.

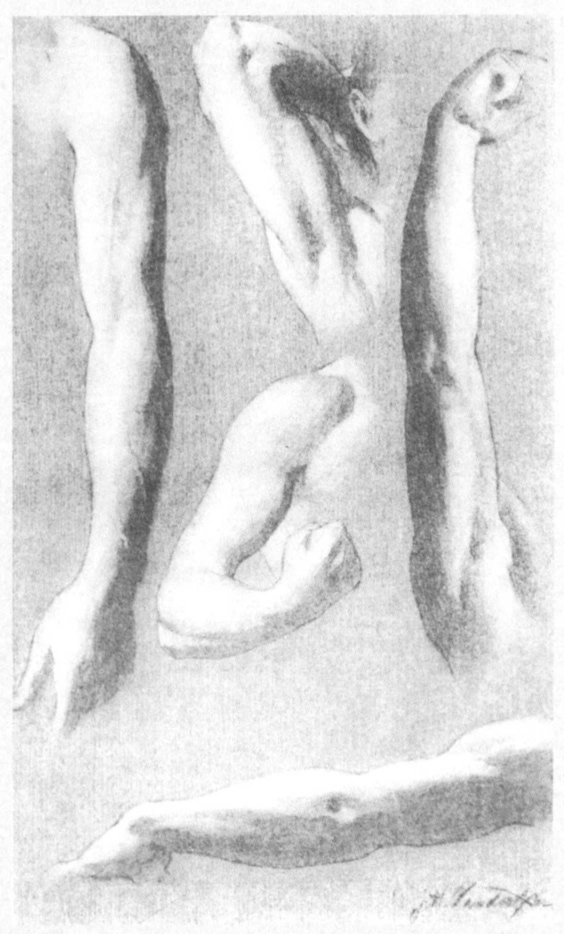

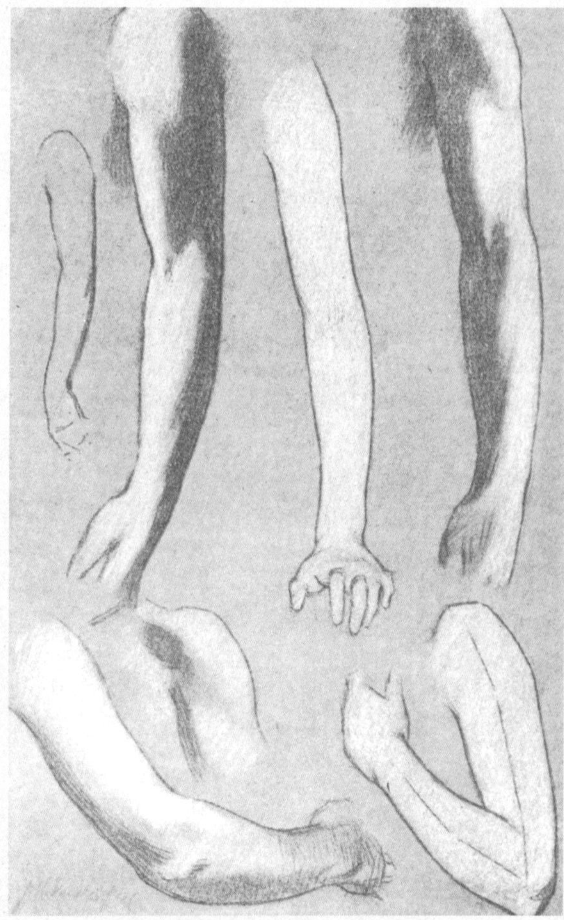

The Hands

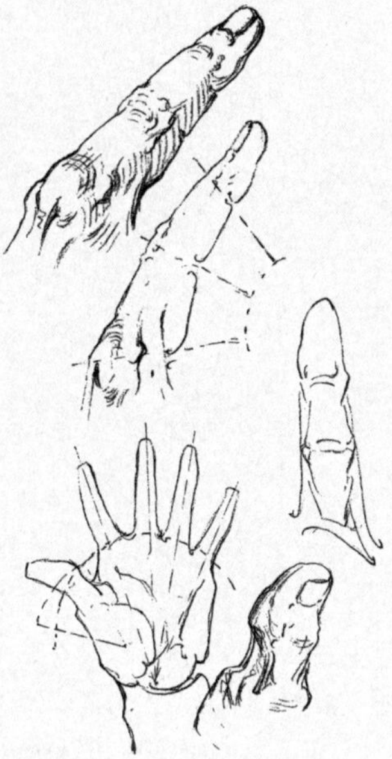

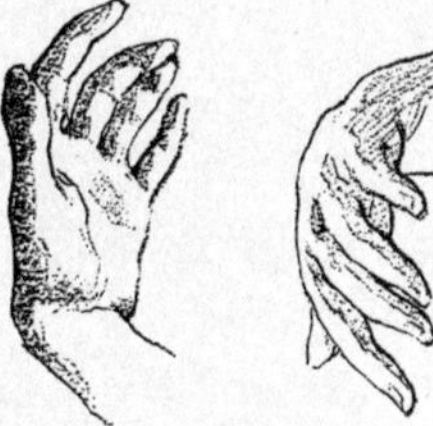

Because hands include so many joints and independently moving parts, they can appear very different depending on the angle of view.

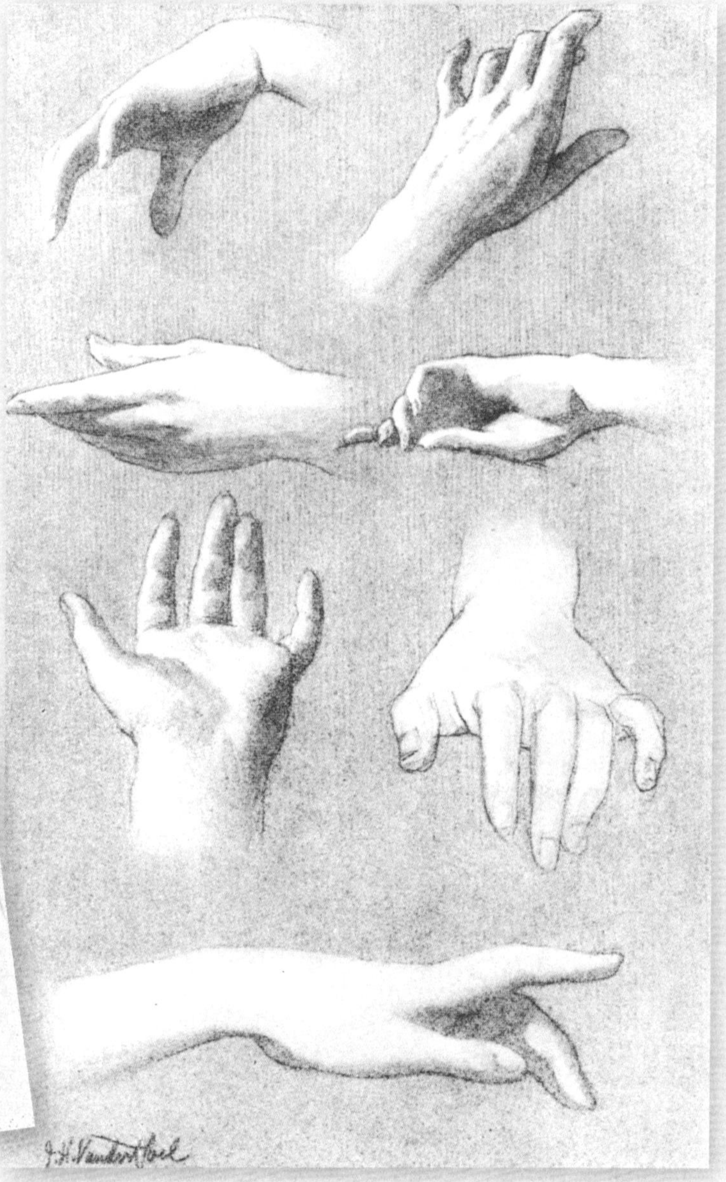

When splayed, fingers have a radial position on the hand. Fingers call for studies of their own, as they are complex structures with multiple joints.

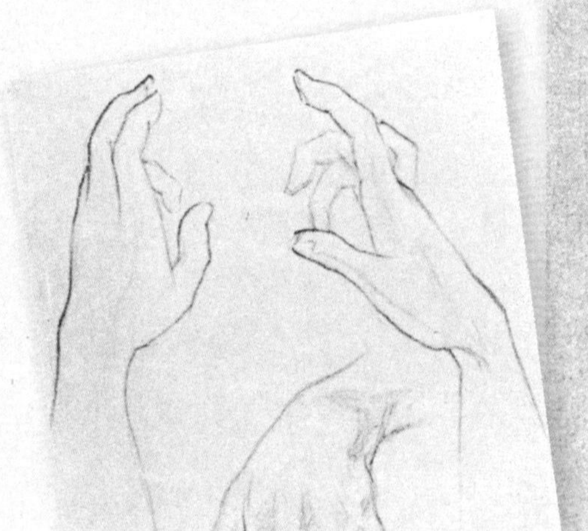

LEGS & FEET

The legs and feet ground the figure and are at the core of both
balance and motion. Studying the musculature and joints is crucial
for creating realistic bodies standing or moving in space.

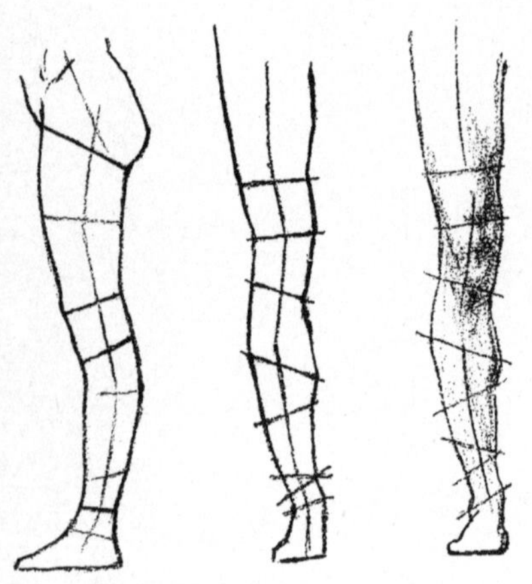

Straight legs aren't actually straight.
From every angle, the flesh and bones
curve. Create maps of all the angles you
can use to define the form of the leg.

Anatomy of the Leg

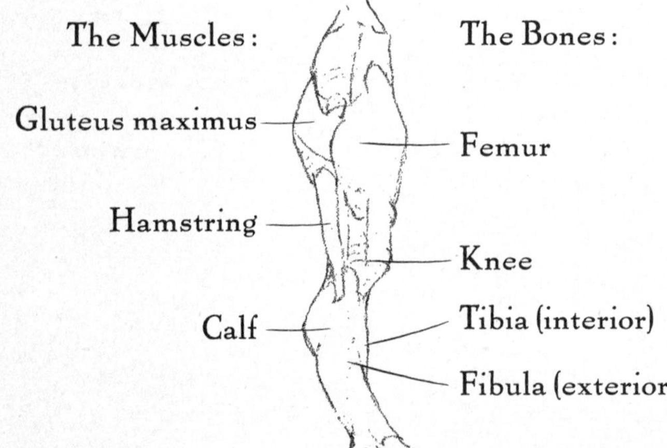

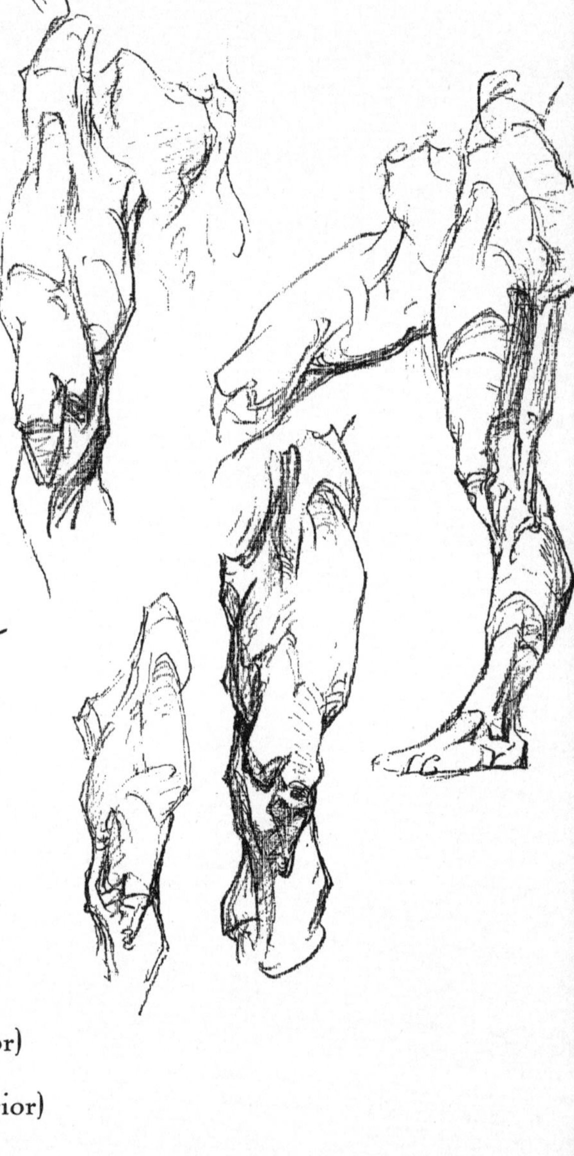

Use the practice pages in this section to draw legs and feet in different positions.

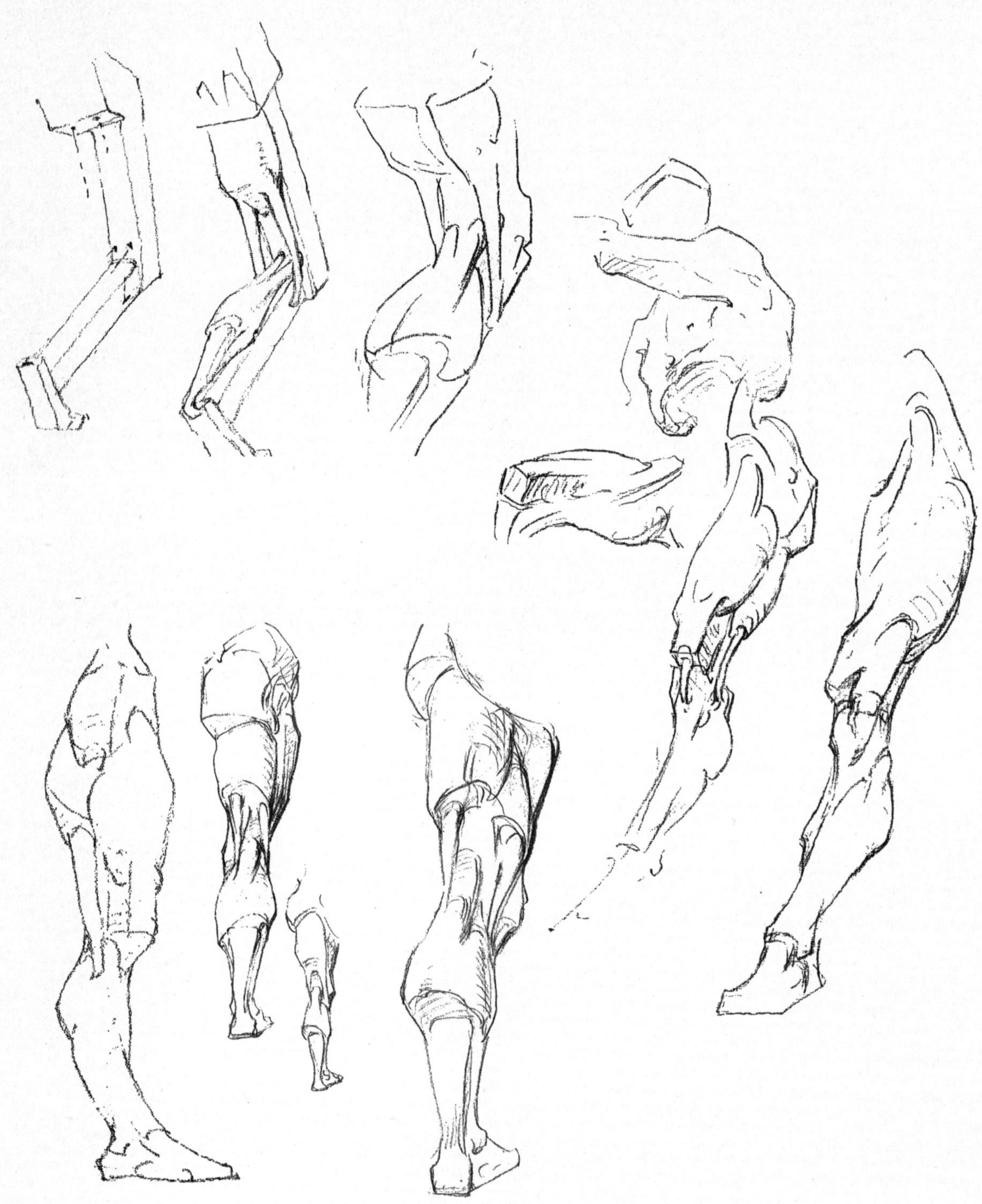

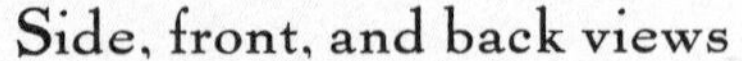

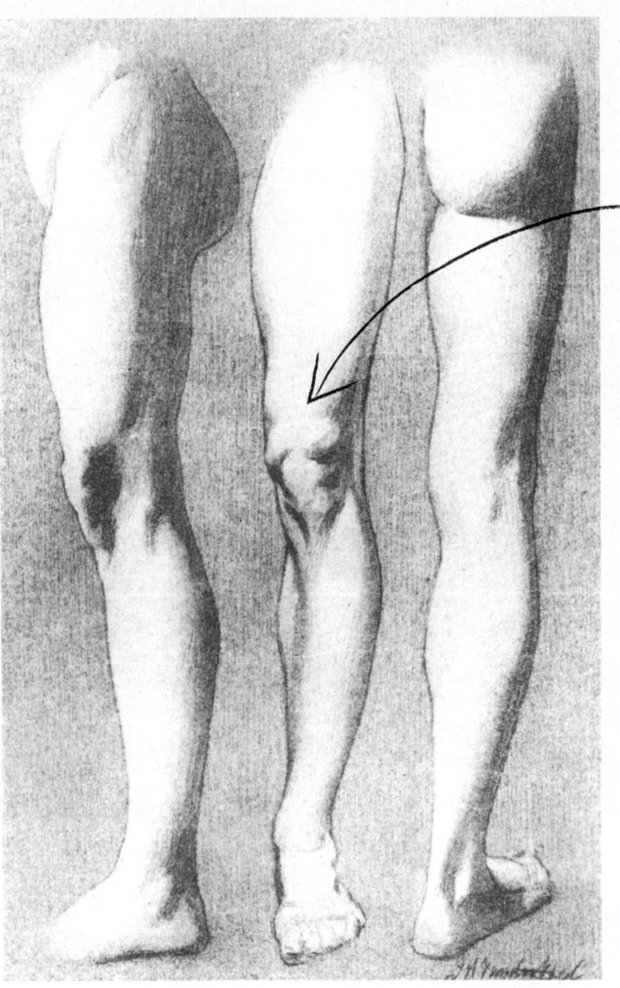

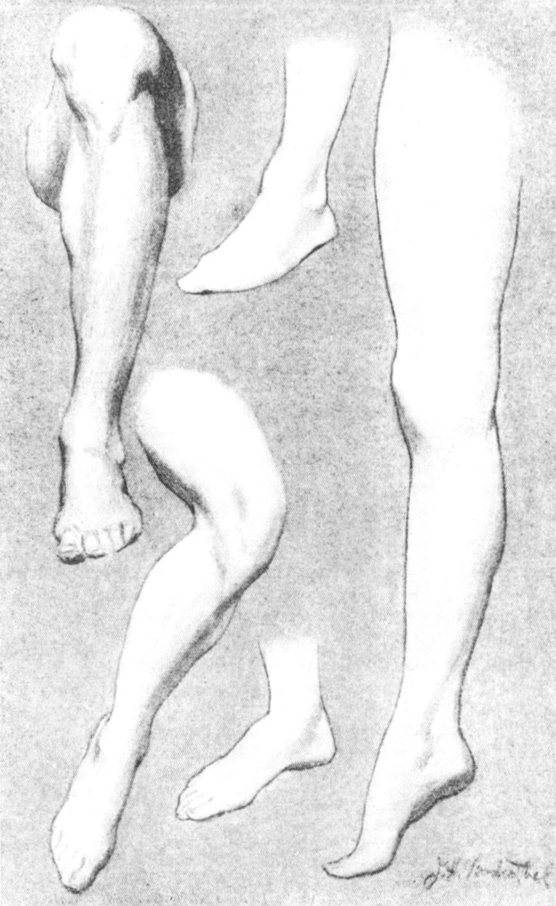

Strong overhead lighting on the knee helps the shadows define the underlying forms.

Study the leg as it bends at different angles and as the model gracefully lifts the heel to point the toes.

The Feet

When blocking in the foot, first establish the angles with
straight lines; then move on to the curves of the arch, ball
of the foot, heel, and toeline.

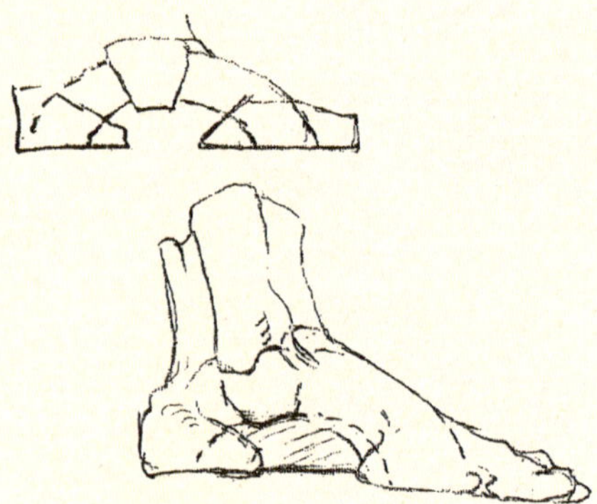

The bones of the foot are wedged
together and bound by ligaments,
giving resistance, solidity, and
elasticity.

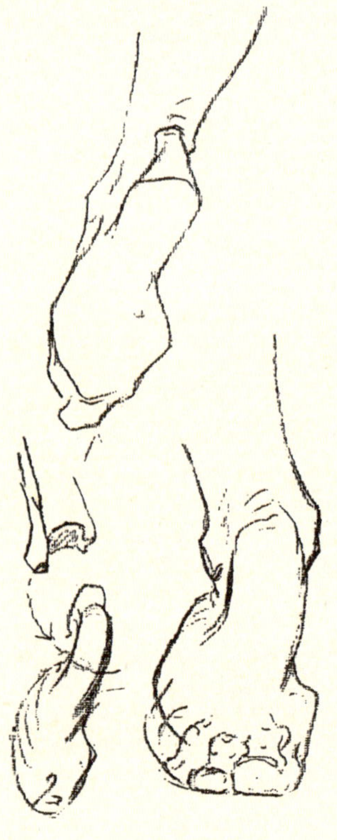

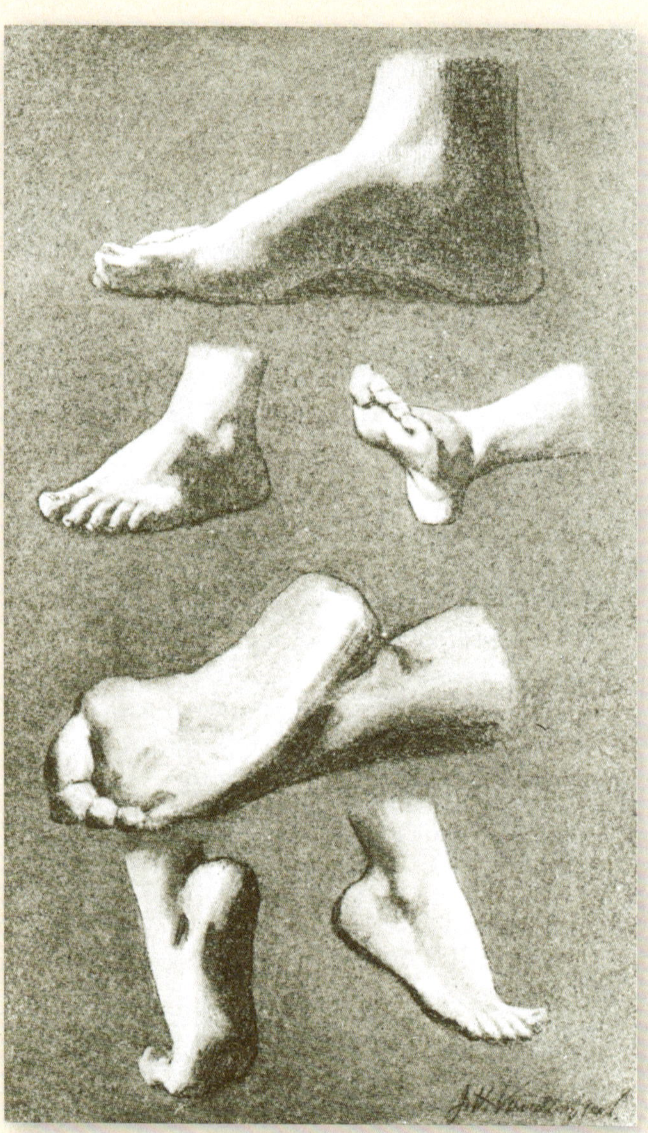

The two bones of the lower leg (the tibia and the
fibula) extend from the knee to the ankle. The lower
ends of these bones project to form the inner and
outer ankle joints, where they receive the articulating
joint of the foot (the astragalus). The foot rolls under
the leg bones.

THE HEAD

Figure drawing generally places an emphasis on the human body. However, the head is a bastion of expression and likeness, and its details should not be excluded from figure drawings. On the following pages, sketch along to study the head's main forms and features.

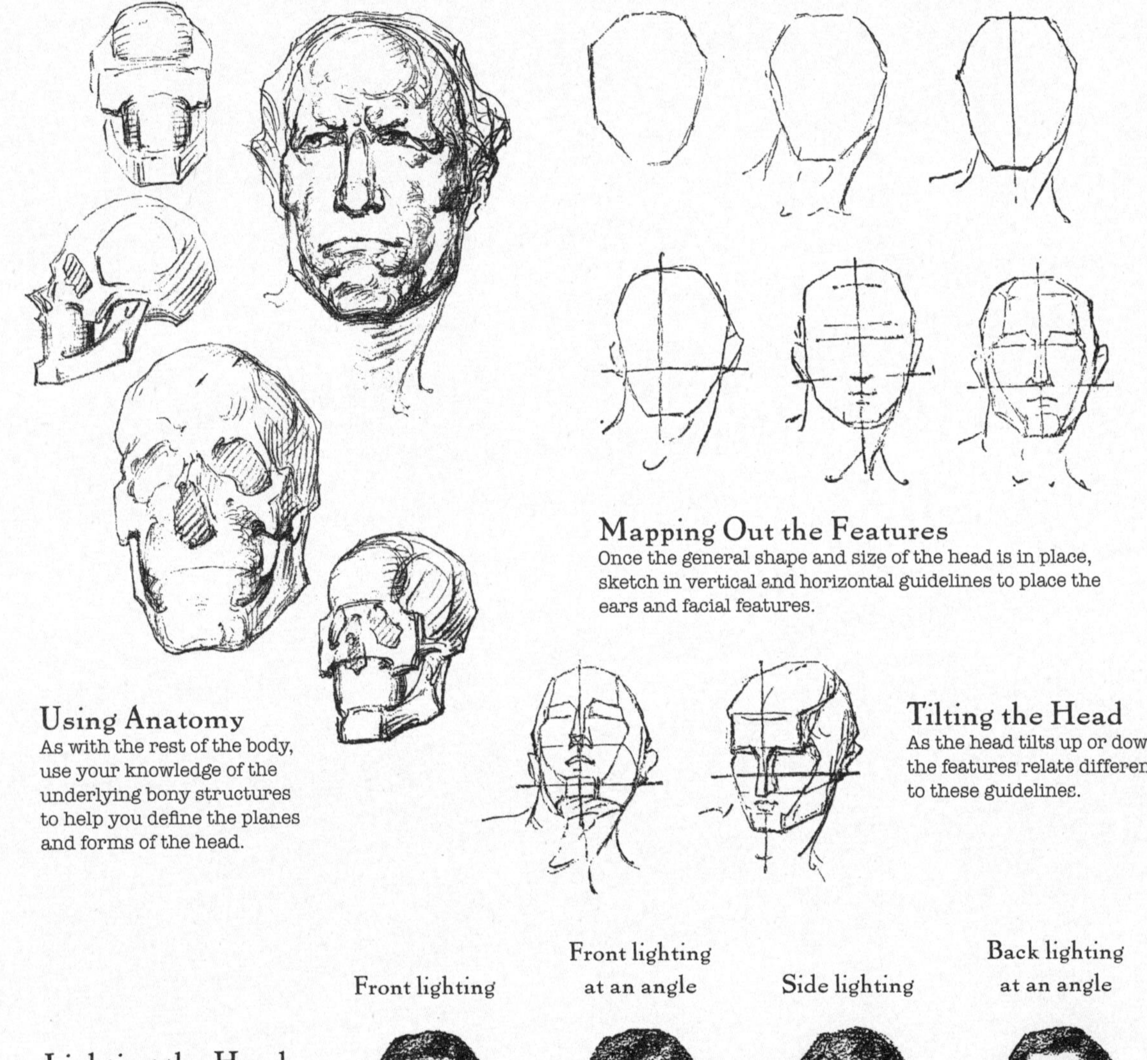

Mapping Out the Features

Once the general shape and size of the head is in place, sketch in vertical and horizontal guidelines to place the ears and facial features.

Using Anatomy

As with the rest of the body, use your knowledge of the underlying bony structures to help you define the planes and forms of the head.

Tilting the Head

As the head tilts up or down, the features relate differently to these guidelines.

Lighting the Head

The strength and direction of lighting can dramatically change the appearance of the head, as well as the mood of the scene.

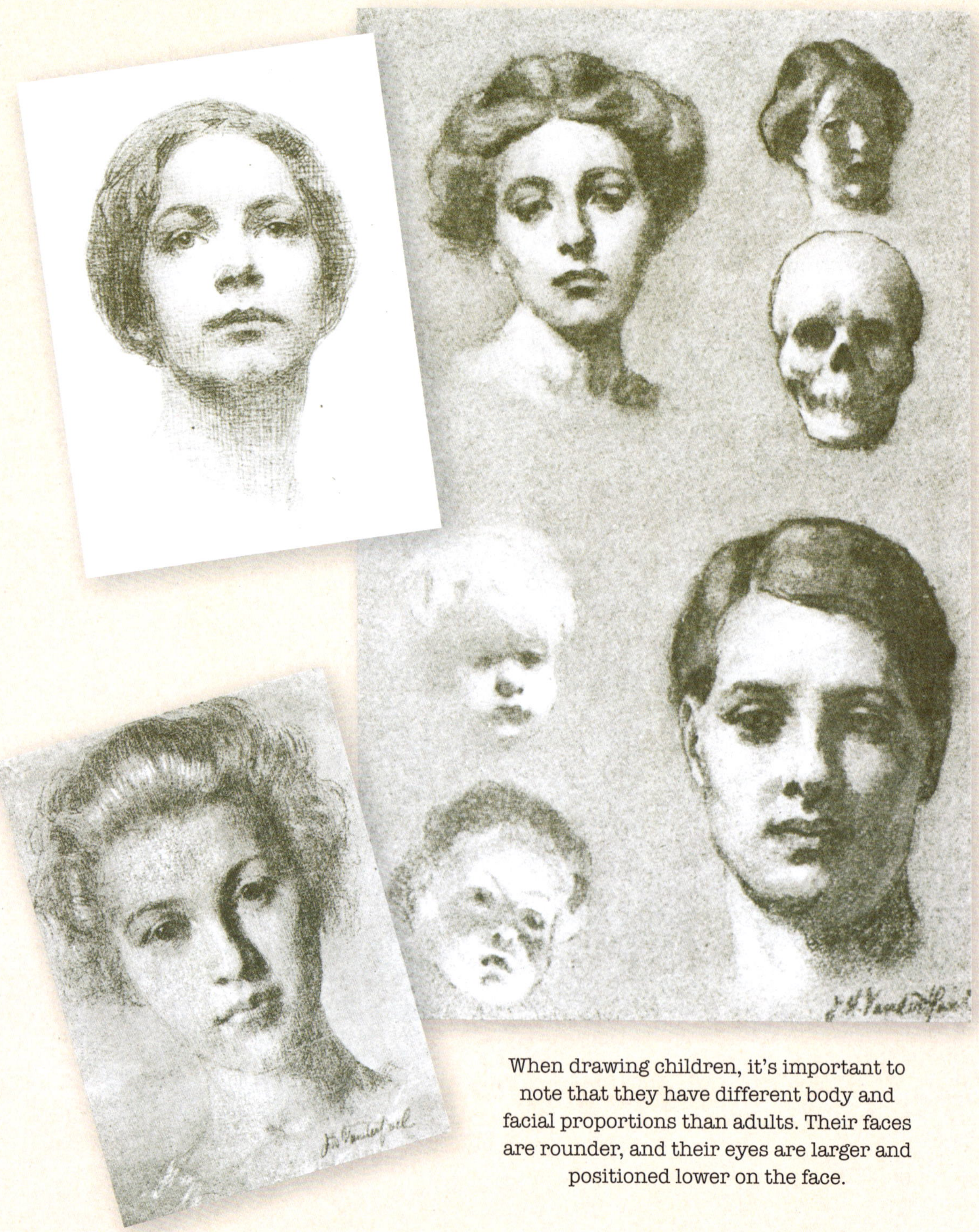

When drawing children, it's important to note that they have different body and facial proportions than adults. Their faces are rounder, and their eyes are larger and positioned lower on the face.

The Neck

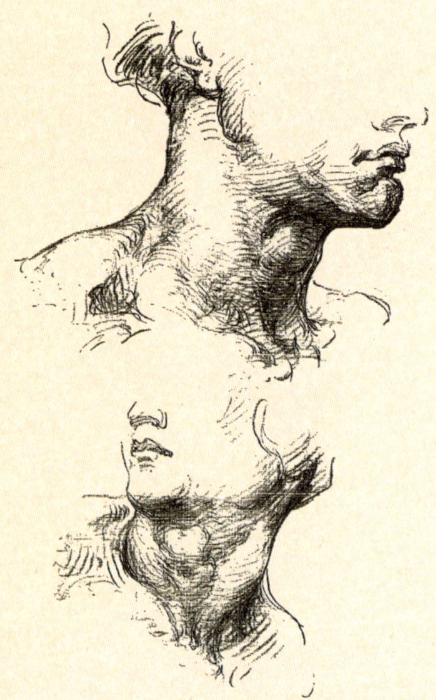

The cylindrical nature of the neck is an excellent subject for practicing cross-contour drawing—a method of shading that uses curved, parallel lines applied in a way that describes the form's surface.

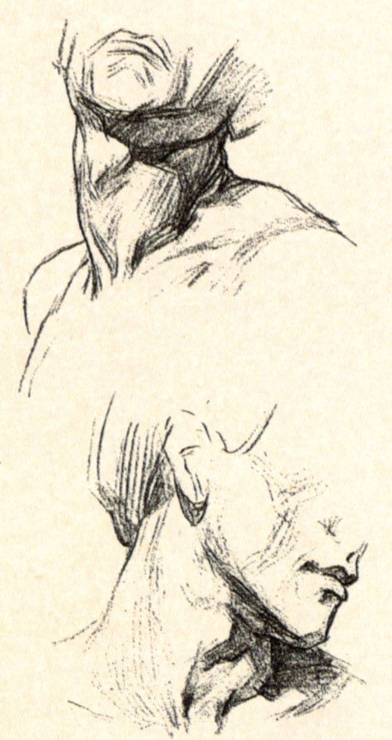

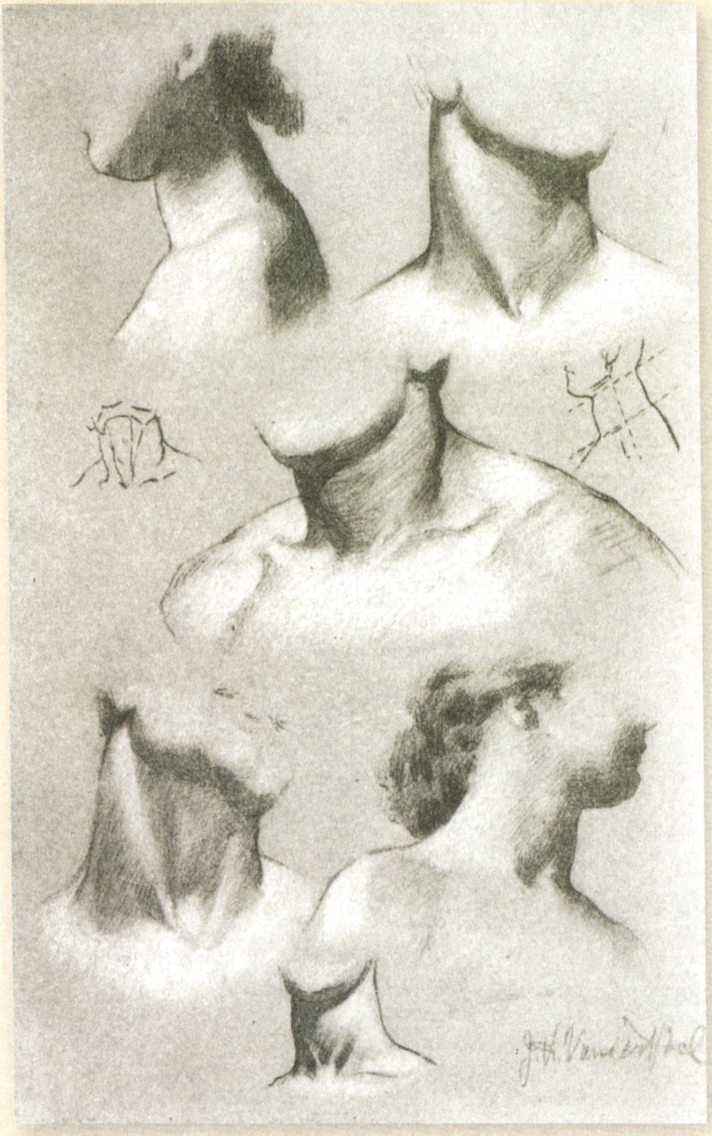

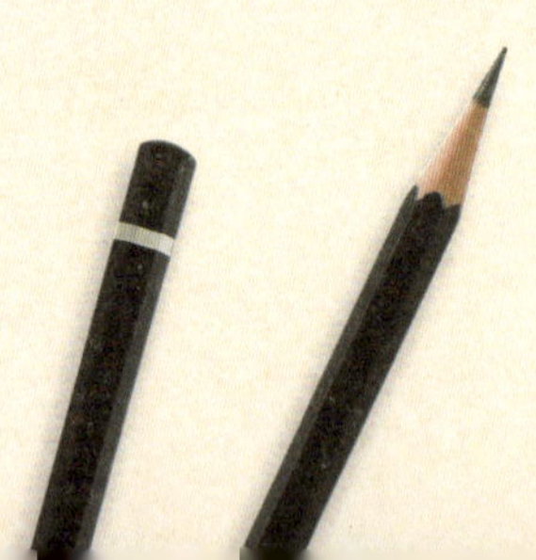

Typically the male neck appears short, thick, and firm, rising from the body vertically. The female neck usually appears long and slender, with a greater forward direction.

The Eyes

The eyes are spherical structures wrapped in the skin of the
eyelids. The concentric circles of the iris and pupil show the
model's direction of sight. Often referred to as the "windows of
the soul," the eyes are key to creating a drawing with a strong
likeness to the model.

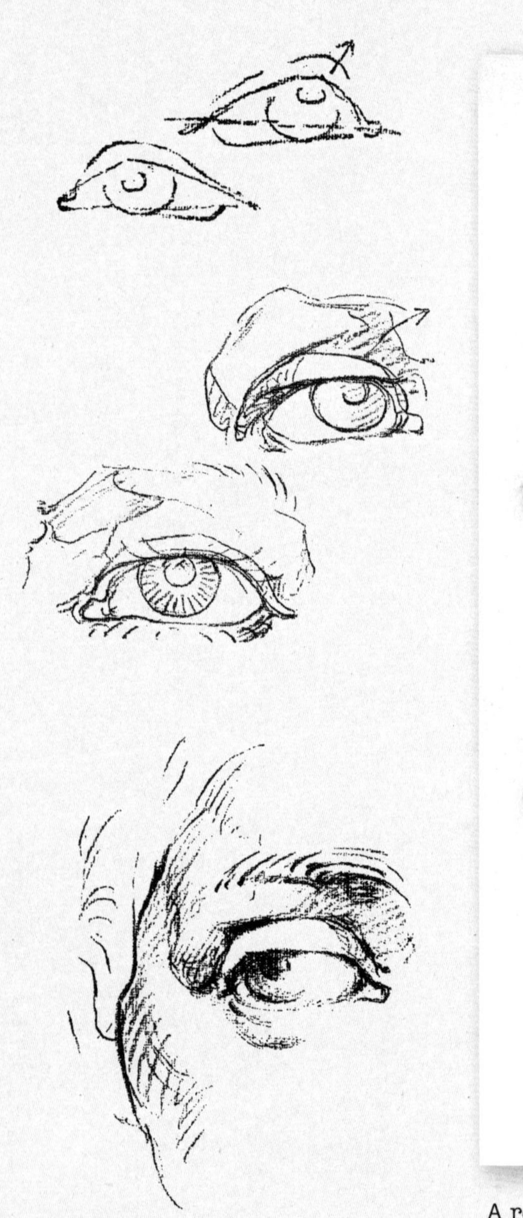

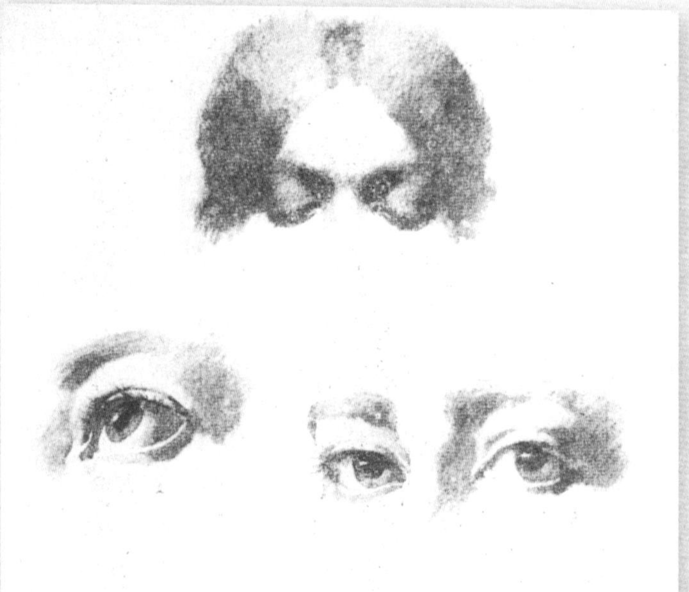

A realistic representation of the eyes relies heavily on the
shadows around it. As you create studies, place the eyes
in context by suggesting the nearby planes and forms.

The Nose

Begin by sketching the nose in relation to the face. Use very simple forms and identify the main planes.

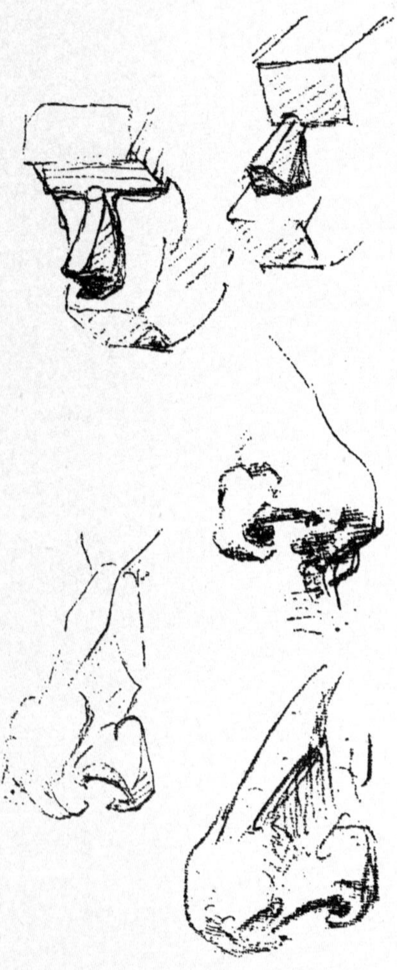

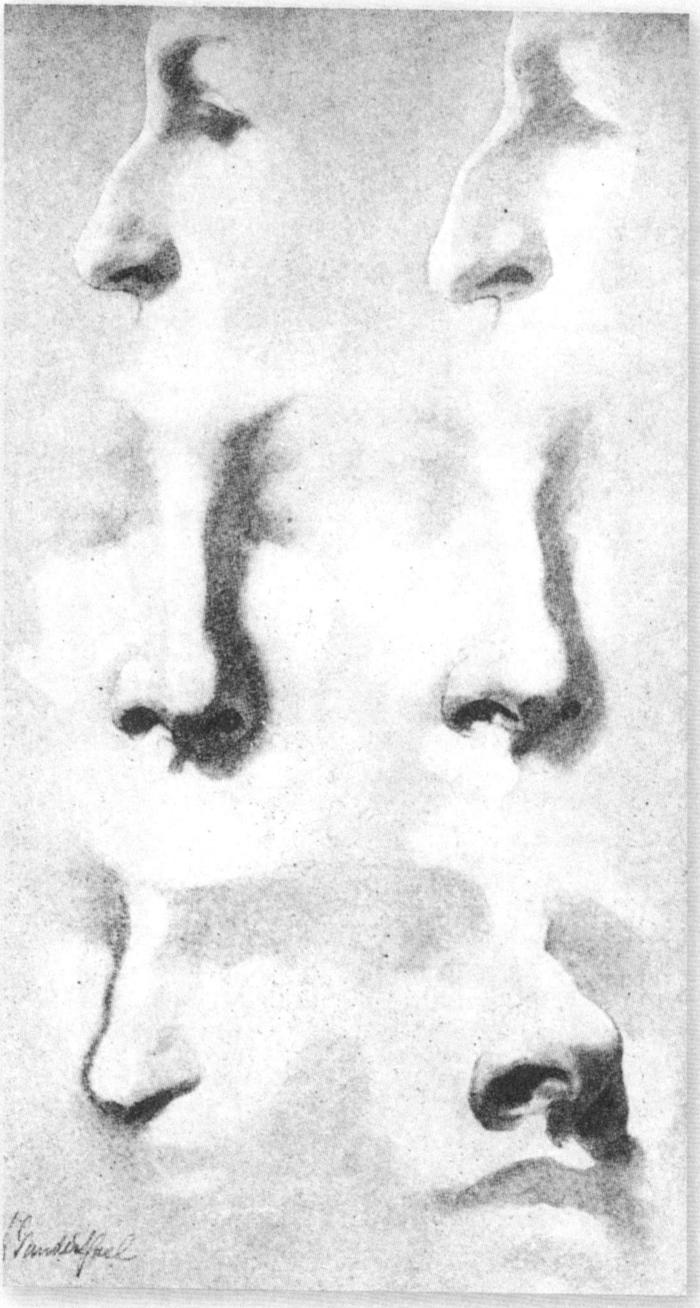

Here are some questions to ask yourself while viewing a model: Does the nose turn up or down? Is the end rounded or chiseled? Is the bridge wide or slim? Does the nose have a straight or curved slope when viewed in profile? What is the shape of the nostril, and how much of it can you see from various viewpoints?

The Ear

The ear may not be considered the most important
feature of the head, but it is surprisingly complex and
challenging to draw. Begin by simplifying the lines and
planes as much as possible.

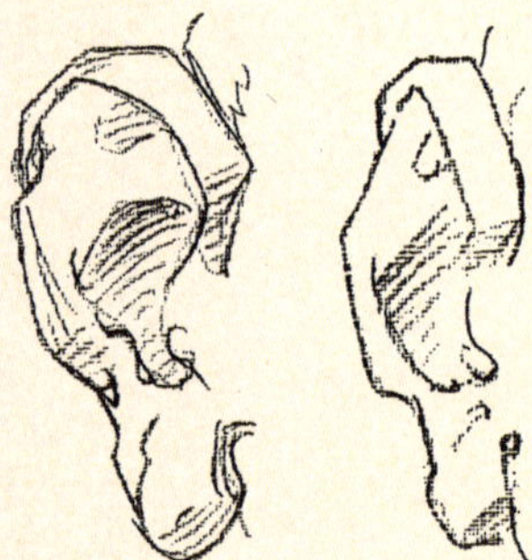

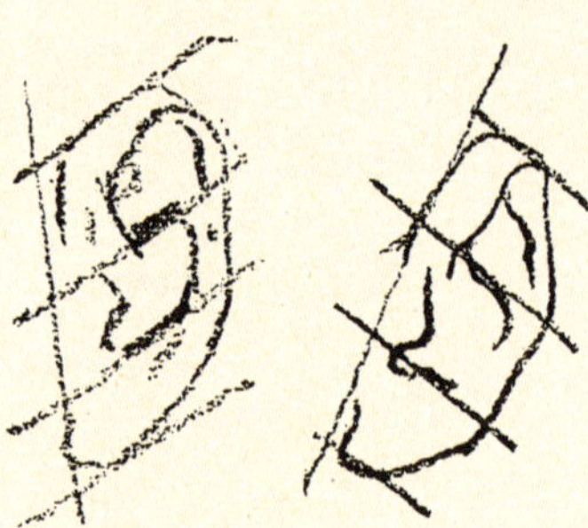

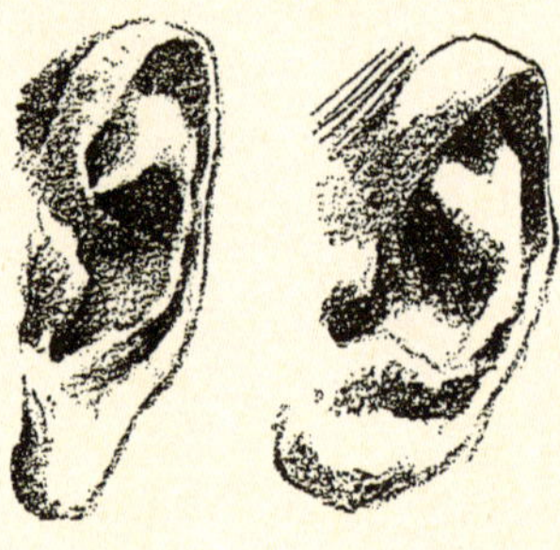

Block in the edges and most
obvious folds; then shade
simply with parallel lines to
indicate planes in shadow.

Some artists view the
ear in three sections as
they establish the initial
propotions and outline.

Ear shapes can vary greatly;
pay close attention to the
folds and lines to achieve a
likeness to the individual.

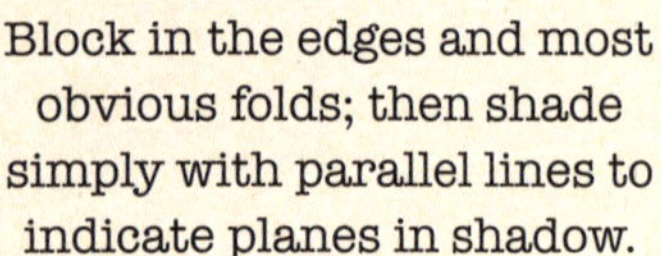

When viewed
from the back, the
ear has an almost
hornlike shape.

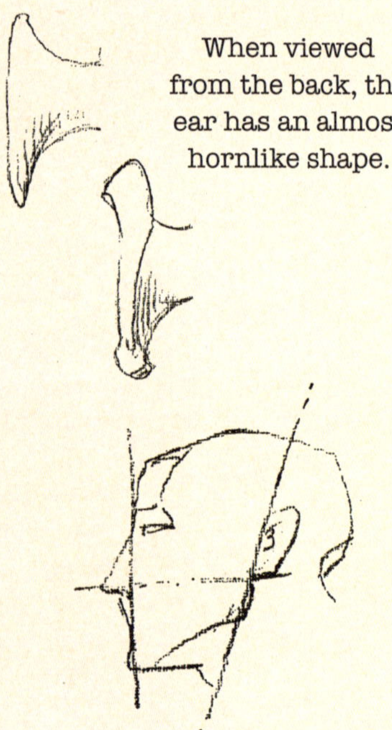

In a side view, note
how the angle of the
ear relates to the
profile of the face.

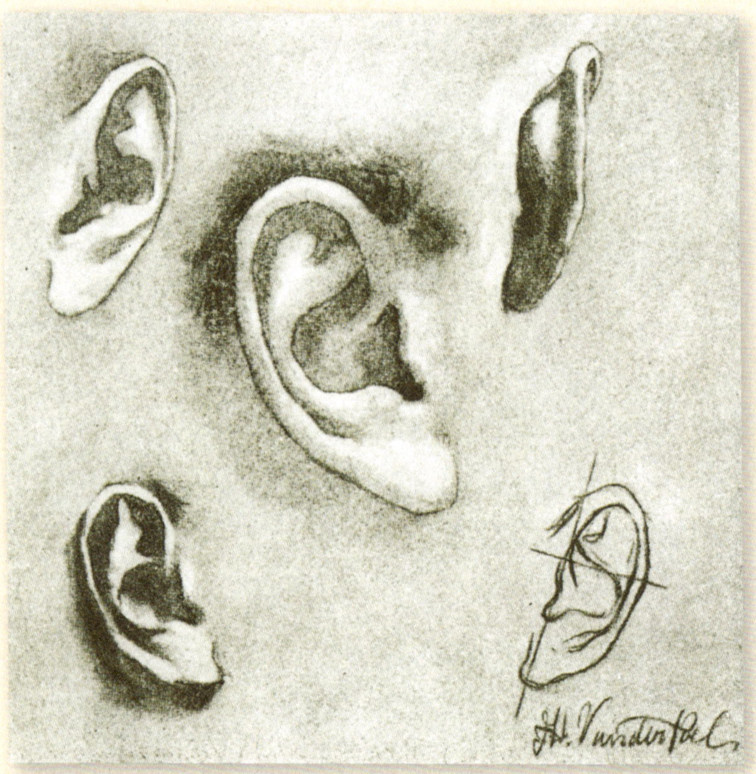

The Mouth

To draw a realistic mouth and lips, artists must
understand how they wrap around the form of the face.
Think of the underlying structures, such as the arch of the
teeth and the cylindrical nature of the midface.

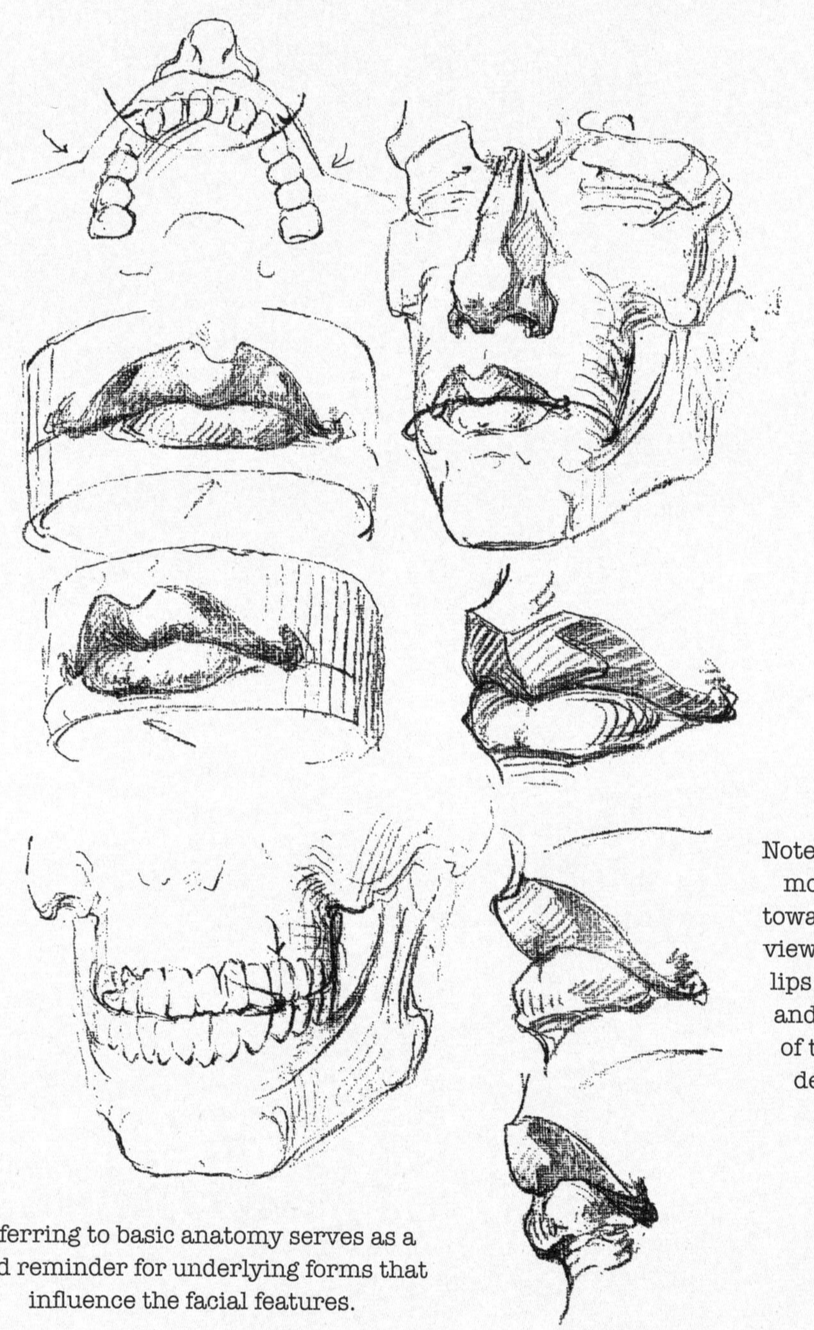

Note that as the
model turns
toward a profile
view, less of the
lips are visible
and the width
of the mouth
decreases.

Referring to basic anatomy serves as a
good reminder for underlying forms that
influence the facial features.

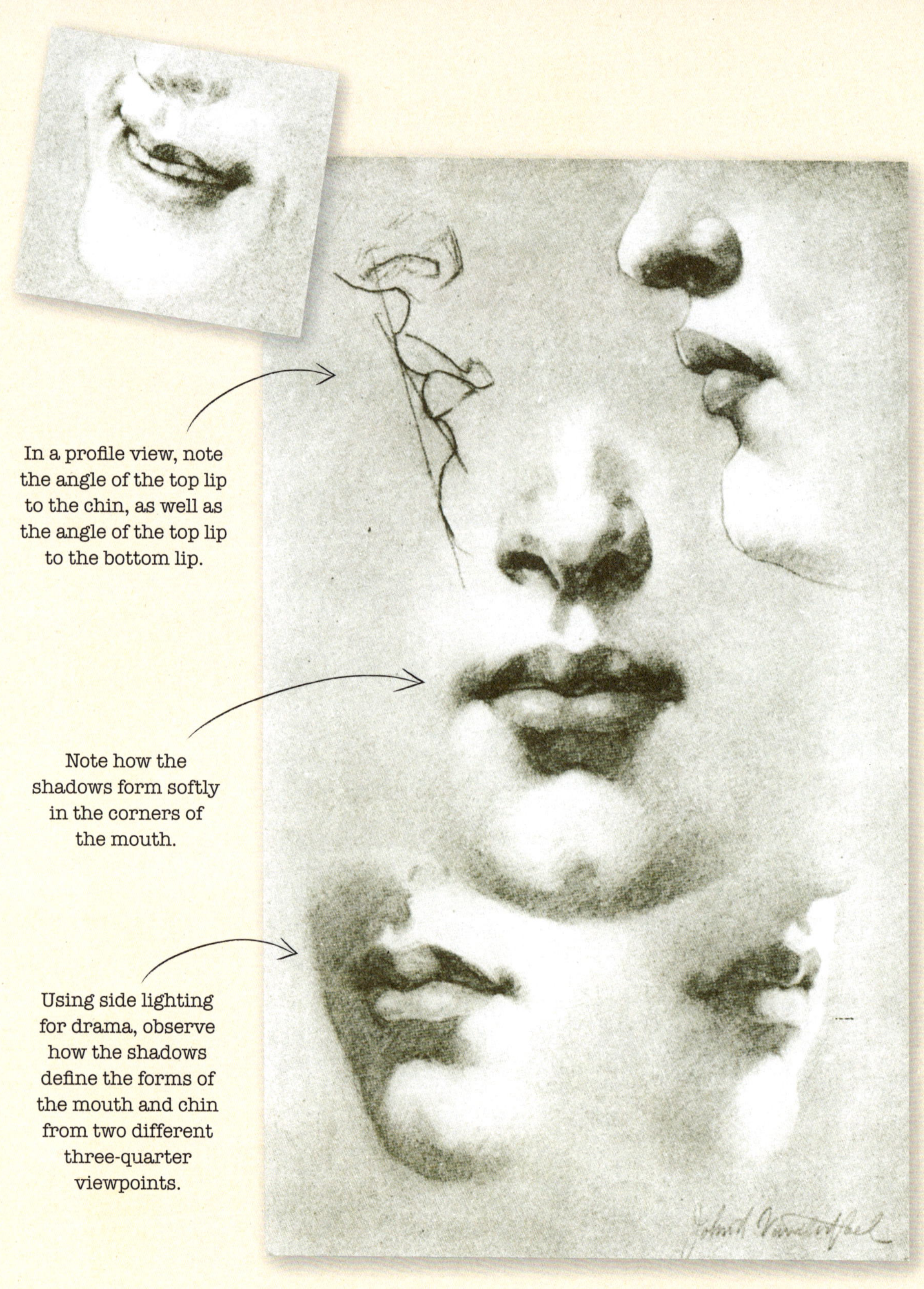

In a profile view, note the angle of the top lip to the chin, as well as the angle of the top lip to the bottom lip.

Note how the shadows form softly in the corners of the mouth.

Using side lighting for drama, observe how the shadows define the forms of the mouth and chin from two different three-quarter viewpoints.

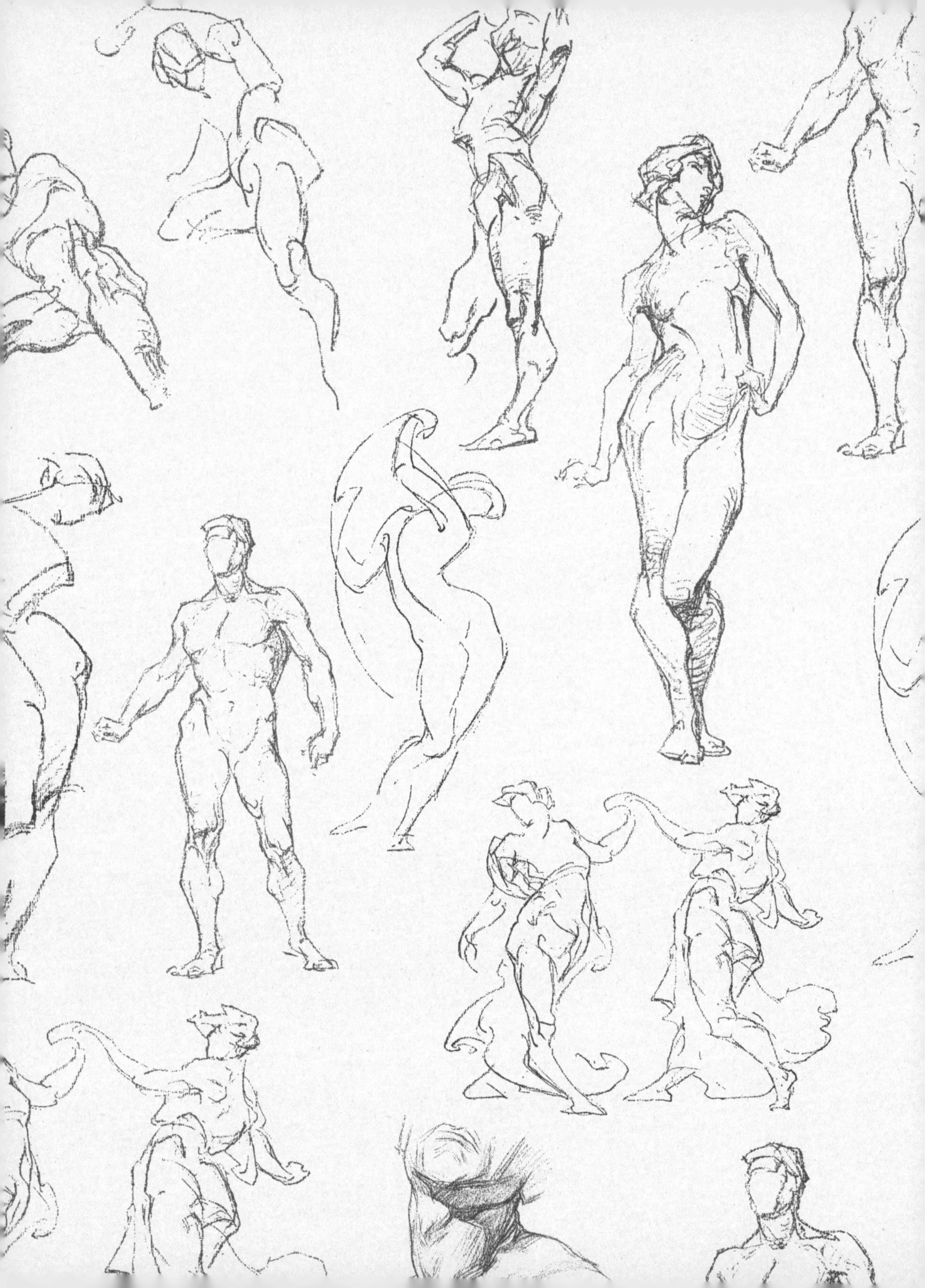

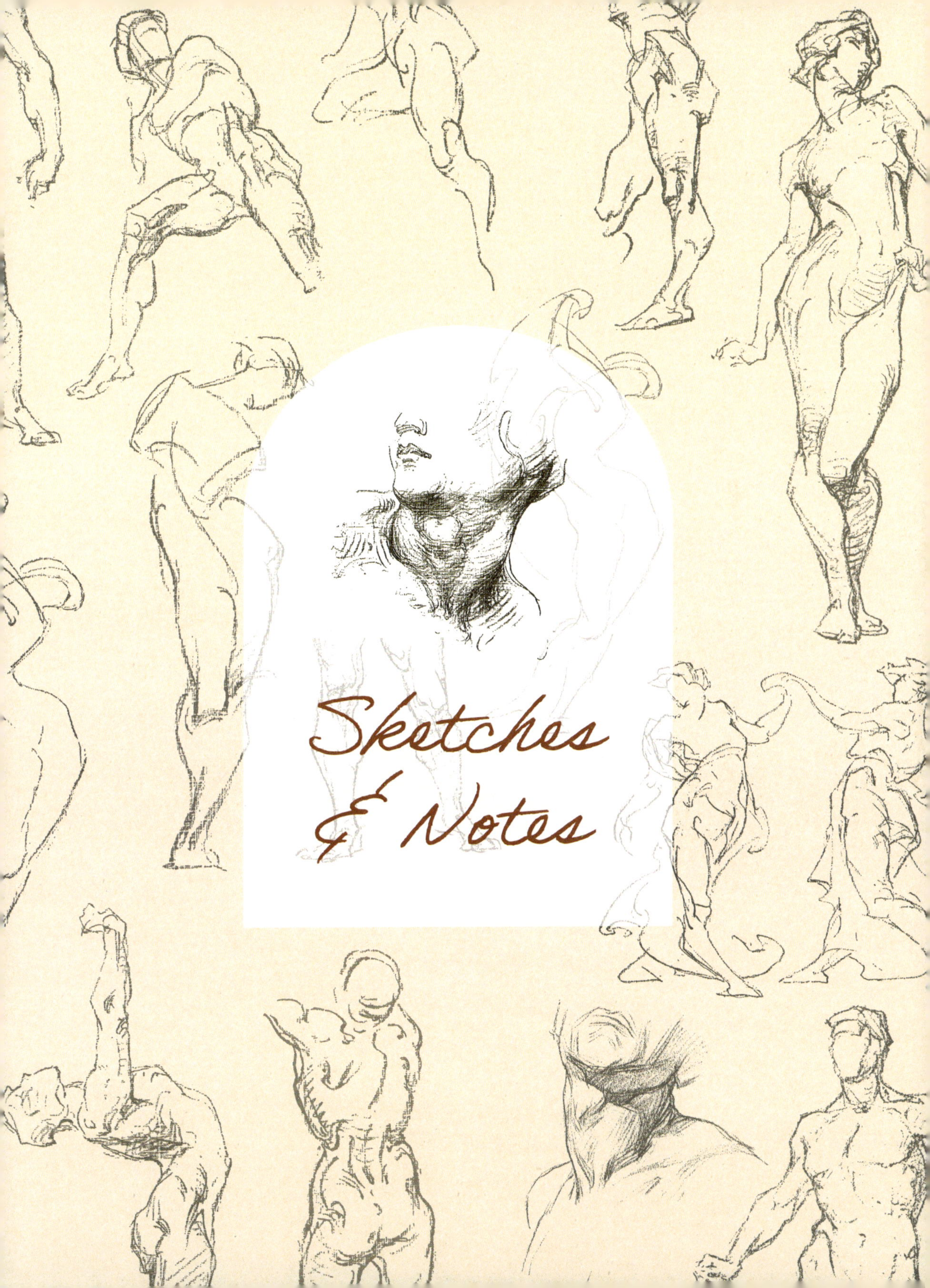
Sketches
& Notes

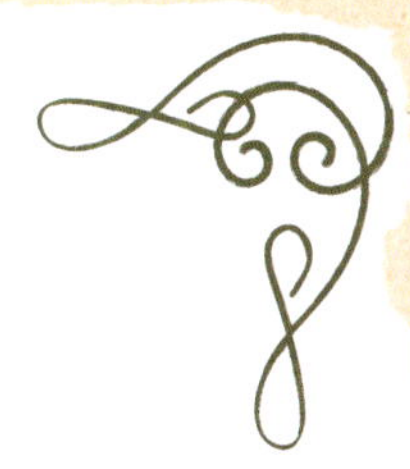

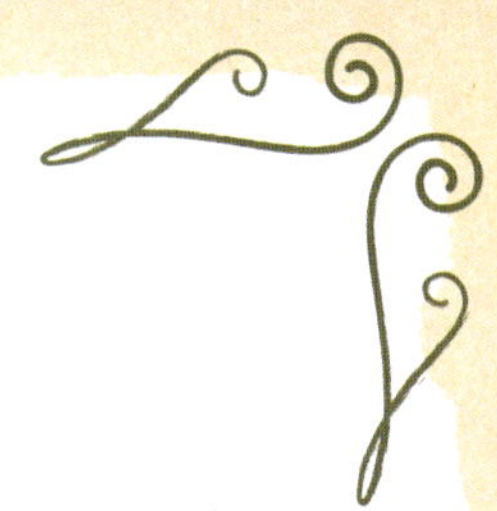

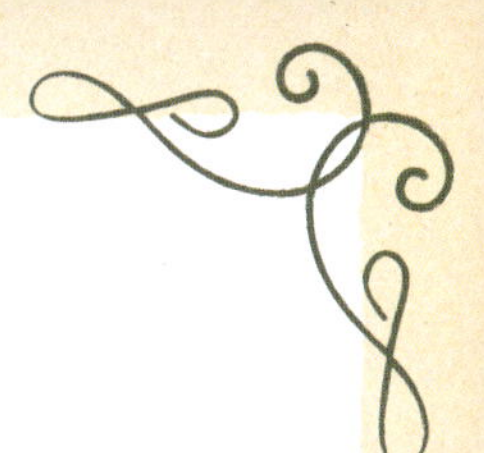

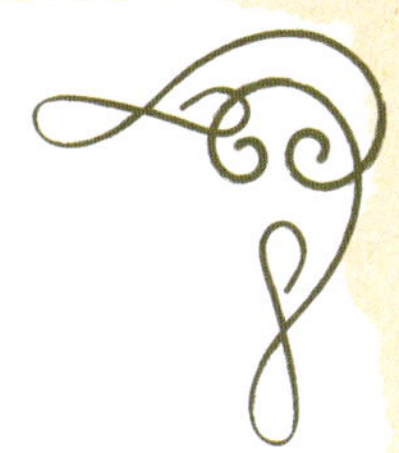

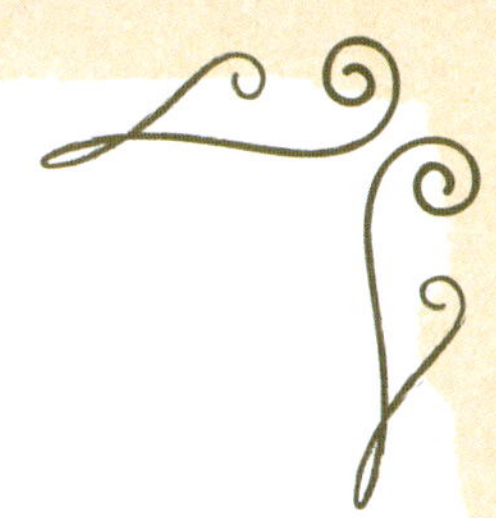